Springer Theses

Recognizing Outstanding Ph.D. Research

AF587592

For further volumes:
http://www.springer.com/series/8790

Aims and Scope

The series "Springer Theses" brings together a selection of the very best Ph.D. theses from around the world and across the physical sciences. Nominated and endorsed by two recognized specialists, each published volume has been selected for its scientific excellence and the high impact of its contents for the pertinent field of research. For greater accessibility to non-specialists, the published versions include an extended introduction, as well as a foreword by the student's supervisor explaining the special relevance of the work for the field. As a whole, the series will provide a valuable resource both for newcomers to the research fields described, and for other scientists seeking detailed background information on special questions. Finally, it provides an accredited documentation of the valuable contributions made by today's younger generation of scientists.

Theses are accepted into the series by invited nomination only and must fulfill all of the following criteria

- They must be written in good English.
- The topic should fall within the confines of Chemistry, Physics, Earth Sciences, Engineering and related interdisciplinary fields such as Materials, Nanoscience, Chemical Engineering, Complex Systems and Biophysics.
- The work reported in the thesis must represent a significant scientific advance.
- If the thesis includes previously published material, permission to reproduce this must be gained from the respective copyright holder.
- They must have been examined and passed during the 12 months prior to nomination.
- Each thesis should include a foreword by the supervisor outlining the significance of its content.
- The theses should have a clearly defined structure including an introduction accessible to scientists not expert in that particular field.

Michael Karl Gerhard Kruse

Extensions to the No-Core Shell Model

Importance-Truncation, Regulators and Reactions

Doctoral Thesis accepted by
the University of Arizona, USA

Author
Dr. Michael Karl Gerhard Kruse
Lawrence Livermore National Laboratory
Livermore, CA
USA

Supervisor
Prof. Dr. Bruce Barrett
University of Arizona
Tucson, AZ
USA

ISSN 2190-5053 ISSN 2190-5061 (electronic)
ISBN 978-3-319-37581-6 ISBN 978-3-319-01393-0 (eBook)
DOI 10.1007/978-3-319-01393-0
Springer Cham Heidelberg New York Dordrecht London

© Springer International Publishing Switzerland 2013
Softcover reprint of the hardcover 1st edition 2013
This work is subject to copyright. All rights are reserved by the Publisher, whether the whole or part of the material is concerned, specifically the rights of translation, reprinting, reuse of illustrations, recitation, broadcasting, reproduction on microfilms or in any other physical way, and transmission or information storage and retrieval, electronic adaptation, computer software, or by similar or dissimilar methodology now known or hereafter developed. Exempted from this legal reservation are brief excerpts in connection with reviews or scholarly analysis or material supplied specifically for the purpose of being entered and executed on a computer system, for exclusive use by the purchaser of the work. Duplication of this publication or parts thereof is permitted only under the provisions of the Copyright Law of the Publisher's location, in its current version, and permission for use must always be obtained from Springer. Permissions for use may be obtained through RightsLink at the Copyright Clearance Center. Violations are liable to prosecution under the respective Copyright Law.
The use of general descriptive names, registered names, trademarks, service marks, etc. in this publication does not imply, even in the absence of a specific statement, that such names are exempt from the relevant protective laws and regulations and therefore free for general use.
While the advice and information in this book are believed to be true and accurate at the date of publication, neither the authors nor the editors nor the publisher can accept any legal responsibility for any errors or omissions that may be made. The publisher makes no warranty, express or implied, with respect to the material contained herein.

Printed on acid-free paper

Springer is part of Springer Science+Business Media (www.springer.com)

To my mother, Ute Kruse

Supervisor's Foreword

Over the last 20 years significant advances have been made in our ability to perform *ab initio* calculations for describing the properties of atomic nuclei. These advances have come in the form of (1) a better understanding of how to construct consistent nucleon–nucleon, three-nucleon, and higher-nucleon interactions based on Quantum Chromodynamics, using Effective Field Theory and Chiral Perturbation Theory; (2) the development of new and/or improved nuclear many-body techniques such as the Green Function Monte Carlo approach, the No Core Shell Model (NCSM) formalism, the Coupled Cluster method, etc., for performing the many-nucleon calculations; and (3) tremendous advances in computer hardware and software, which allow for much larger and more complicated numerical calculations to be performed.

The principal question facing this area of research is how to extend the successes for these new nuclear many-body methods to heavier-mass nuclei beyond mass number $A = 16$, e.g., ^{16}O. The problem has to do with the exponential growth of the model spaces required in the numerical calculations, in order to obtain converged results, as the number of nucleons in a nucleus is increased. Present computing technology cannot handle these huge dimensions, which are far greater than the existing limit of about 10^{10} configurations. A great deal of time and effort is presently being devoted to this problem. Some of these new approaches are (1) the *ab initio* Shell Model with a Core, (2) the Importance Truncation method, (3) the NCSM in an EFT Framework, (4) the Monte Carlo-NCSM, (5) the In-Medium Similarity Renormalization Group, (6) the Symmetry-adapted NCSM [SP(3, R)], and a number of other techniques. This Ph.D. thesis has to do with the second of these methods.

One of the most promising approaches for attacking this problem is to develop a physically motivated way for truncating these giant model spaces to manageable sizes, which can be handled by present-day computers, while also retaining the states essential for capturing all the underlying physics of the nucleus being investigated. Considerable effort is currently being invested in this challenge, with the Importance Truncation method being one of the leading candidates under development. The main idea is to truncate the number of configurations kept in the basis space, based on an importance criterium for first-order-in-perturbation-theory additions to a reference wave function, such as that for the ground state, as the size of the basis space is increased in order to achieve a converged result. The details of this procedure are clearly presented and discussed in Dr. Kruse's Ph.D. thesis. In his dissertation research, Dr. Kruse not only greatly improved the existing Importance Truncation formalism, by using a sequential method for adding new configurations, but also studied in detail how to quantify the size of the theoretical error in the final extrapolated results. Such an error quantification is essential for understanding the accuracy and reliability of the results obtained by the Importance Truncation formalism, as applied to the NCSM and the No Core Shell Model/Resonating Group Method (NCSM/RGM) approaches for nuclear structure and reaction calculations, respectively.

Besides the significant work on the IT-NCSM, this thesis also includes new results on the application of the IT-NCSM to ^{9}He within the NCSM/RGM and an important discussion and analysis of the Ultraviolet (UV) and Infrared (IR) limits for large model-space calculations. In particular, the work in the thesis found that a specific choice of an IR regulator leads to a scaling behavior for calculated ground-state energies, provided one has captured all the UV physics in the underlying nuclear interaction. Such an insight places extrapolations of large model-space calculations on a solid footing.

To summarized, the findings in Dr. Kruse's Ph.D. thesis have made a very important contribution to our understanding of the Importance Truncated-NCSM (or IT-NCSM) method and its ability to produce physically motivated truncated model spaces for performing NCSM and NCSM/RGM calculations for nuclei beyond mass $A = 16$ nuclei. This is a highly significant advancement for the application of microscopic nuclear many-body techniques to heavier-mass nuclei.

Acknowledgments

There are a great number of people who have either contributed directly or indirectly to make this thesis possible. To my advisor, Bruce Barrett, it has been a great honor to work with you on a daily basis. Your patience, insights, and support have given me the tools to freely pursue exciting research in my own way; I cannot express my gratitude to having been your student. To Petr Navrátil, for believing in my abilities, for all the encouragement, and for being a constant source of optimism. To Sid Coon, who treated me as a colleague, taught me the value of clearly expressing ideas and for being a mentor. I will miss our conversations. To Hank Miller, in emphasizing the importance of seeing the big picture, and for all his interesting 'cute' physics problems. To Alexander Lisetskiy, for getting me started with NCSM calculations and high-performance computing.

Sean Fleming and Bira van Kolck, for all the discussions we had over lunch. The folks at LLNL, Eric Jurgenson, Sofia Quaglioni, Erich Ormand and Jutta Escher. I would particularly like to thank Eric for discussing programmatic issues with the IT-NCSM code as well as thank Sofia for being so patient while I learned the NCSM/RGM formalism from her.

Walter Meyer at the University of Pretoria, for his teaching and for his guidance. Johann-Heinrich Schoenfeldt, for all the lessons of staying true to oneself (Godspeed).

To my high school science teachers, Mr. Cruickshanks and Mrs. Koster, for teaching in a way that inspired me to explore the mysteries of science.

To my fellow classmates, Ian, Michael, Gregg, Elliott, Jiamin, Sybil, Tim and Brian, for making the experience of grad-school a whole lot better.

To Swati Singh, for being a constant source of inspiration, for the times I needed motivation, for being a best friend. Your laughter will be missed dearly.

To my parents, who came from a modest background, for the years they worked long hours so that I have a better life. I will always remember the importance of hard work.

To Sybil, my wife; for all the testing times we survived in grad-school (and there were many), your unwavering support was always there. Tucson will forever be our home.

Contents

Chapter 1
Introduction to Low-Energy Nuclear Physics

1.1 An Overview for Non-Experts

The No-Core Shell Model (NCSM) is a first-principles nuclear structure technique, with which one can calculate the observable properties of light nuclei ($A \leq 20$). It is considered ab-initio as the only input to the calculation is the nuclear Hamiltonian, which contains realistic two or three-nucleon (NN or NNN) interactions. Provided the calculation is performed in a large enough basis space, the ground-state energy will converge. For $A \leq 4$, convergence has been demonstrated explicitly. The NCSM calculations are computationally very expensive for $A > 6$, since the required basis size for convergence often approaches on the order of a billion many-body basis states. In this thesis we present three extensions to the NCSM that allow us to perform larger calculations, specifically for the p-shell nuclei. The Importance-Truncated NCSM, IT-NCSM, formulated on arguments of multi-configurational perturbation theory, selects a small set of basis states from the initially large basis space, in which the Hamiltonian is now diagonalized. Previous IT-NCSM calculations have proven reliable, however, there has been no thorough investigation of the inherent error in the truncated IT-NCSM calculations. We provide a detailed study of IT-NCSM calculations and compare them to full NCSM calculations in an attempt to judge the accuracy of IT-NCSM in heavier nuclei. Even when IT-NCSM calculations are performed, one often needs to extrapolate the ground-state energy from the finite basis (or model) spaces to the infinite model space. Such a procedure is commonplace but does not necessarily have the ultraviolet (UV) or infrared (IR) physics under control. We present a potentially promising method that maps the NCSM parameters into an effective-field theory framework, in which the UV and IR physics is treated appropriately. The NCSM is well suited to describing bound-state properties of nuclei, but is not well adapted to describe loosely bound systems, such as the exotic nuclei near the neutron drip line. With the inclusion of the resonating group method (RGM), the NCSM/RGM can provide a first-principles description of exotic nuclei. The NCSM/RGM is also the first extension of the NCSM that can describe dynamic processes such as nuclear reactions.

M. K. G. Kruse, *Extensions to the No-Core Shell Model*,
Springer Theses, DOI: 10.1007/978-3-319-01393-0_1,
© Springer International Publishing Switzerland 2013

1.2 From QCD to Nuclear-Structure

It has been a long-standing goal of the nuclear physics community to describe the properties of all nuclei from their fundamental interactions [1]. Unlike atomic systems, in which the underlying fundamental interaction is the Coulomb force, which is rather well-understood, nuclei and their nuclear interactions are far more complicated, displaying a rich set of phenomena. A quick survey of this rich behavior is, for example, the deuteron (the only two-nucleon bound system), the recently discovered halo nuclei (*e.g.*, ^{6}He), collective modes in heavier nuclei, and dynamic properties, such as α- and β-decay.

In Quantum Electrodynamics (QED), the fundamental interaction between light and matter, properties, such as the Lamb-Shift in the Hydrogen atom, can be calculated from a perturbative expansion of the various possible interactions, each order of the expansion being smaller than the previous one. This framework is possible, since the coupling constant of the interactions, $\alpha \approx \frac{1}{137}$, is small, and each order of the perturbation is, in effect, proportional to a higher power in α than the previous order. In hadronic systems, the fundamental interaction is known as Quantum Chromodynamics (QCD). Quarks, the constituents of hadrons, such as protons or neutrons, interact with each other via gluons, the force carriers of QCD. Once again, there is a coupling constant present in the theory, most commonly denoted as α_s. However, the value of α_s depends on the momentum scale of interest; at large momentum scales, α_s tends to be small, but at low momentum scales, it grows rapidly. The reference momentum to be used, here in determining which regime applies to a particular system, is referred to as $\Lambda_{QCD} \sim$ 1,000 Mev/c. This dependence of the coupling constant on the momentum under consideration is known as the running of the coupling constant. In the case of low-energy nuclear physics, the typical momentum scales are much smaller than Λ_{QCD}, which means that α_s is actually quite large. This places us in a difficult situation. If we want to describe nuclei, using QCD as the fundamental interaction, we need to specify at least how two nucleons interact with each other, before we proceed any farther. Yet, a perturbative expansion of QCD is not possible in that regime, due to the running of α_s. The consequences of this statement and its resolution will be discussed later.

The nuclear landscape has roughly 270 stable nuclides, about 2,000 known nuclei, and perhaps as many as another 3,000 presently unknown nuclei. Describing all of these nuclei directly from QCD is a daunting challenge and is presently simply not feasible. As any reasonable physicist will argue, the degrees of freedom that describe a single nucleon, the deuteron, or those of light nuclei, or even heavier systems that exhibit collective dynamics are all different, even though they must all originate from QCD at some level. The question on how to unify all phenomena into one, single framework, at the heart of which lies QCD, is, and will be for a long time, an open question. On the other hand, many-body physicists view this hurdle as simply one of fine detail, which will be overcome, and are generally satisfied to work with an interaction, usually based on other degrees of freedom (and thus, simpler physics). Such a stance has often led to the expression that many-body practitioners work with

'God given' potentials. That is not to say that they do not appreciate the work that has been done to connect low-energy nuclear physics to QCD, they simply view the more interesting part of the problem to be the many-body aspect. As has been mentioned, there are many known nuclei, each with its own specific degrees of freedom and properties. Some very clever techniques have been developed to handle most parts of the nuclear landscape, from the *ab-inito* nuclear-structure techniques of light nuclei (the focus of this dissertation) to traditional shell-model calculations of mid-mass nuclei and density-functional theory for heavy nuclei.

1.3 Theoretical Nuclear-Structure of Light Nuclei

The main area of interest of this dissertation will be the calculation of the nuclear-structure of light nuclei. For these nuclei, we can use recently developed *ab-initio* techniques. By that, we literally mean calculations that are based on minimal assumptions of the underlying physics. In fact, all we assume is that the nuclear interactions used are in some way directly connected to QCD.

One would assume that, once the interaction between two-, or possibly more, nucleons is specified, solving the many-body problem is simply a matter of turning a large mathematical crank. In essence, all one has to do is solve the Schroedinger equation. Such a statement greatly understates the true nature of any many-body calculation, even though one really *does* solve the Schroedinger equation. What is often a simple procedure for two interacting particles can become a computational nightmare for A interacting particles. The real question is, how does one deal with all these interacting particles in a sensible and tractable manner? To begin with, one needs to specify the nature of the interaction among the nucleons. Typically, interactions involve only two or also three nucleons. However, there could, in principle, be as many as A-body interactions. The need for A-body interactions is however somewhat artificial, considering that it is generally believed that four-or higher-body interactions are incredibly small in size, given the success of calculations that include only two-and three-body forces. Next, one would like to specify a convenient basis in which to work. In principle one can choose any basis, but computationally some choices are better than others (although from a physics point, perhaps is a poorer choice, such as choosing the harmonic oscillator (HO) basis over the Woods-Saxon basis). Often, one is faced with a basis in which one or several difficulties can arise: there are many basis states to handle, or the antisymmetrization is particularly difficult, or the basis has the incorrect asymptotic features of the physical problem, just to name a few issues. Each many-body technique must face this difficulty, and often the rewards of a particular choice come with a price in another aspect of the calculation. One tries to remedy these issues as best as possible. At this stage, one needs to turn to powerful computers to perform the calculation, most often the actual diagonalization of the Hamiltonian, which is yet another interesting aspect of the problem. The computational facets of the problem have, in fact, themselves become quite relevant. These details will be discussed in the next chapter.

1.4 *Ab-Initio* Techniques of Nuclear-Structure

Several ab-initio techniques have been developed, in order to calculate the properties of light nuclei. The most common ones, that deal with $A > 4$, are the No Core Shell model (NCSM) [2–4], the Green's-function Monte-Carlo technique (GFMC) [5–8], and the Coupled Cluster technique with singles and doubles (CCSD) [9, 10]. Each technique has it's own features. The No Core shell model is the closest to a traditional shell-model calculation, except that all nucleons are active in the model-space and realistic interactions are employed. Unfortunately, the basis is difficult to handle for $A > 16$. The Green's function technique uses a Monte-Carlo approach to sample the basis states, but unfortunately can only deal with local coordinate space interactions (and most also deal with the Monte Carlo sign problem). The Coupled-cluster technique can extend calculations to much heavier systems than the previously stated two methods, but is only useful in the vicinity of doubly-magic nuclei. There are other methods too, more suitable for $A \leq 4$, such as the Faddeev [11] and Faddeev-Yakubovsky techniques [12–14], or the hyperspherical harmonic oscillator (HH) techniques [15, 16], or the NCSM expressed in Jacobi coordinates [17]. All of these techniques predict binding energies, that are in agreement with the experimental binding energies of light nuclei, such as the Triton or the alpha-particle, provided realistic interactions are used [18]. One of the major advancements of our understanding of nuclear-structure, lie in the calculations that include a three-body force. Without the inclusion of the three-body force, nuclei are underbound in the theoretical calculations, when compared to experimental data. In slightly heavier systems, $4 \leq A \leq 16$, the calculations performed with the No Core Shell Model or Green's function Monte-Carlo, provide reasonably good agreement with experiment [8, 19], but are computationally much more challenging. The current implementations of these two techniques, given the computational resources available, reach their limit in this mass range. If one wants to reach fully converged results, or attempt to do heavier systems, one needs to augment these techniques either through some sensible modification, or through extrapolation to fully converged results. The latter is currently not feasible when $A > 20$, so one must rethink the approach.

1.5 Extensions to the NCSM

The No Core shell model has been demonstrated to calculate the properties of mostly stable, light nuclei, up to $A \leq 20$ quite successfully. Beyond that mass range, or for loosely bound nuclei, the method runs into computational difficulties. In order to reach any sort of meaningful results, or ideally make some predictive calculations, one must employ large model-spaces (i.e., a large basis space). Large model-spaces imply that a large number of basis states must be stored in computer memory (RAM), which is very limited, considering all the other data that must be stored. One could opt to use a more suitable basis, but after roughly 15 years of computer code development, and hard work, one would prefer to exploit the existing technology and gains. There are other physics related issues, when employing a different basis, which will be discussed in the chapter on the No Core Shell model. On the other hand, what if one can use the No Core shell model, retaining all of its capabilities, with perhaps some minor modifications, or by extending the model to treat various systems to overcome some of these hurdles? This is the main question I will address in this thesis.

1.5.1 Importance-Truncated NCSM

As has been mentioned in the previous paragraph, the large number of basis states thwarts the calculations of heavier nuclei, and also those of halo nuclei. One might be tempted to ask if one really needs all of these basis states. If one could develop a method for pre-selecting some basis states considered relevant to the calculation, and at the same time discard a large number of seemingly irrelevant basis states, then the calculations could be done rather simply. This is what the Importance-Truncated No Core Shell model (IT-NCSM) formulation does [20, 21]. Although the method works very well, allowing for some previously inaccessible calculations to be done, one might wonder how much information has been discarded, when some basis states are excluded from the calculation. The formulation of the IT-NCSM, how it is typically used, the effect on observables as well as the criticisms of the method, will be discussed in Chap. 3.

1.5.2 The Extrapolations to an Infinite Model-Space

NCSM calculations in which $A > 4$ converge slowly with the size of the basis and thus need to be extrapolated to the infinite basis space. There are a couple of different extrapolation techniques in use today [22, 23], however, they are based purely on experiences of past calculations. We will present an extrapolation technique based on the ideas of an effective field theory, in which we remap the parameters of the NCSM onto well defined ultraviolet (UV) and infrared (IR) regulators. These regulators are then used to extrapolate to the infinite basis. We now give a brief overview.

When one solves the nuclear Hamiltonian in the NCSM framework (i.e., using the HO basis), the Hamiltonian assumes a dependence on the chosen HO energy ($\hbar\Omega$). This unfortunately leads to the results, such as the ground state (gs) energy, being dependent on $\hbar\Omega$ for a given model-space. In practice, this dependence is almost constant, for a range of chosen oscillator energies. Most many-body physicists will argue that one must choose a value of $\hbar\Omega$, that minimizes the gs energy, or, at the very least, choose a value for which the dependence is roughly constant. From an effective-theory point of view, this is rather upsetting. Initially, one began with a Hamiltonian that does not depend on a property of the underlying basis, such as the unperturbed level spacing; however, the final results do depend on it. Admittedly, in the limit of an infinite model-space, this dependence disappears for all values of $\hbar\Omega$, but in practical cases (i.e., finite model-spaces), there are still visible remnants of it. In Chap. 4, we perform a thorough investigation of the effective theory properties of the NCSM, in which we try to ultimately remove the dependence on $\hbar\Omega$. Furthermore, the analysis is done not through single-particle properties, but instead through UV and IR regulators, as is usually done in effective field theories. Recently, there has been some discussion on what the appropriate IR regulator should be. Our analysis will also discuss this important point.

1.5.3 The NCSM/RGM and Exotic Nuclei

Exotic nuclei, which, for example, could be near the neutron-drip lines, or are perhaps only resonances, cannot be calculated by the NCSM accurately. The difficulty lies in the underlying basis of HO wavefunctions, since, these wavefunctions have the incorrect asymptotic behavior. In order to reach convergence, one has to use a large number of oscillator basis states, leading once again to the difficulties of handling a large number of basis states. Even if one could use a procedure like the IT-NCSM, one would only be able to go so far. The NCSM on its own calculates only bound-state properties, making the treatment of resonance states impossible (in its present form), or for that matter, any scattering properties. There are however other techniques, which deal with the continuum properties of nuclei, one of them being the resonating group method (RGM) [24, 25]. When the NCSM basis is coupled with the RGM basis, one can treat loosely bound nuclei, and can also calculate scattering quantities (such as resonances). This recent development, has brought the realm of nuclear reactions on a truly ab-inito footing. The formalism of the NCSM/RGM technique, as well as an application to the ^{9}He system, will be discussed in Chap. 5.

1.6 The Community of Low-Energy Nuclear Physics

Before concluding the introduction, I would like to draw attention to one more relevant discussion. Low-energy nuclear physics has seen quite a vigorous revival in the last twenty years. The revival itself has been a conscious effort by the nuclear physics community, as is mostly laid out in its long-range plans [1]. However, individuals in the community can only make so much progress on their own. In the United States, there has been some infrastructure put in place to bring the community together in order to continue the progress that has been made. The Institute for Nuclear Theory (INT) [26], hosted at the University of Washington, began roughly twenty years ago, and hosts a wide variety of programs, that discuss certain areas of nuclear physics on a regular basis. In fact, it was fit at the INT, under the program *Effective Field Theories and the Many-Body Problem* in 2009, that I was presented with my main thesis topic, Importance-truncation in the No-Core shell model. In the last five years, the creation of UNEDF [27], a program designed to create a 'universal nuclear energy density functional', has brought almost the entire theoretical nuclear-structure community together. The program has also realized that important contributions can be made from other areas of science, namely computer science and mathematics, which are already bearing fruit in the form of some of the most advanced computer codes for investigating nuclear-structure [28, 29]. Funding agencies, such as the Department of Energy (DOE) and the National Science Foundation (NSF) have created these programs and supported them, including the required computational resources. Some particularly notable awards are, for example SciDAC (Scientific discovery through Advanced Computing) [30] as well as INCITE (Innovative & Novel Computational Impact on Theory and Experiment) [31]. On the experimental side, facilities around the world, such as those found at GSI in Germany, and RIKEN in Japan, have supplied the community with new and exciting results. The proposed US facility, the Facility for Rare Isotope Beams (F-RIB) [32], which Congress has funded at 550 million dollars (as of 2012), will continue the current drive and ambitions of the nuclear community, towards new discoveries in this exciting field.

References

1. Nuclear Science Advisory Committee, The Frontiers of nuclear science, a long range plan (2008), arXiv:0809.3137 [nucl-ex]
2. P. Navrátil, B.R. Barrett, Phys. Rev. C **57**, 562 (1998), http://link.aps.org/doi/10.1103/PhysRevC.57.562
3. P. Navrátil, J.P. Vary, B.R. Barrett, Phys. Rev. Lett. **84**, 5728 (2000), http://link.aps.org/doi/10.1103/PhysRevLett.84.5728
4. P. Navrátil, S. Quaglioni, I. Stetcu, B.R. Barrett, J. Phys. G: Nucl. Part. Phys. **36**, 083101 (2009), http://stacks.iop.org/0954-3899/36/i=8/a=083101
5. B.S. Pudliner, V.R. Pandharipande, J. Carlson, S.C. Pieper, R.B. Wiringa, Phys. Rev. C **56**, 1720 (1997), http://link.aps.org/doi/10.1103/PhysRevC.56.1720
6. R.B. Wiringa, S.C. Pieper, J. Carlson, V.R. Pandharipande, Phys. Rev. C **62**, 014001 (2000), http://link.aps.org/doi/10.1103/PhysRevC.62.014001
7. S.C. Pieper, R.B. Wiringa, Annu. Rev. Nucl. Part. Sci. **51**, 53 (2001), http://www.annualreviews.org/doi/pdf/10.1146/annurev.nucl.51.101701.132506, http://www.annualreviews.org/doi/abs/10.1146/annurev.nucl.51.101701.132506
8. S.C. Pieper, K. Varga, R.B. Wiringa, Phys. Rev. C **66**, 044310 (2002), http://link.aps.org/doi/10.1103/PhysRevC.66.044310
9. R.J. Bartlett, M. Musiał, Rev. Mod. Phys. **79**, 291 (2007), http://link.aps.org/doi/10.1103/RevModPhys.79.291
10. G. Hagen, T. Papenbrock, D.J. Dean, M. Hjorth-Jensen, Phys. Rev. Lett. **101**, 092502 (2008), http://link.aps.org/doi/10.1103/PhysRevLett.101.092502
11. L.D. Faddeev, Zh Eksp. Teor. Fiz. **39**, 1459 (1960)
12. O.A. Yakubovsky, Sov. J. Nucl. Phys. **5**, 937 (1967)
13. W. Glöckle, H. Kamada, Phys. Rev. Lett. **71**, 971 (1993), http://link.aps.org/doi/10.1103/PhysRevLett.71.971
14. F. Ciesielski, J. Carbonell, Phys. Rev. C **58**, 58 (1998), http://link.aps.org/doi/10.1103/PhysRevC.58.58
15. M. Viviani, A. Kievsky, S. Rosati, Few-Body Syst. **18**, 25 (1995) ISSN 0177-7963, 10.1007/s006010050002, http://dx.doi.org/10.1007/s006010050002

16. N. Barnea, W. Leidemann, G. Orlandini, Nucl. Phys. A **650**, 427 (1999), ISSN 0375-9474, http://www.sciencedirect.com/science/article/pii/S037594749900113X
17. P. Navrátil, G.P. Kamuntavičius, B.R. Barrett, Phys. Rev. C **61**, 044001 (2000), http://link.aps.org/doi/10.1103/PhysRevC.61.044001
18. H. Kamada, A. Nogga, W. Glöckle, E. Hiyama, M. Kamimura, K. Varga, Y. Suzuki, M. Viviani, A. Kievsky, S. Rosati, et al., Phys. Rev. C **64**, 044001 (2001), http://link.aps.org/doi/10.1103/PhysRevC.64.044001
19. P. Navrátil, V.G. Gueorguiev, J.P. Vary, W.E. Ormand, A. Nogga, Phys. Rev. Lett. **99**, 042501 (2007), http://link.aps.org/doi/10.1103/PhysRevLett.99.042501
20. R. Roth, P. Navrátil, Phys. Rev. Lett. **99**, 092501 (2007), http://link.aps.org/doi/10.1103/PhysRevLett.99.092501
21. R. Roth, Phys. Rev. C **79**, 064324 (2009), http://link.aps.org/doi/10.1103/PhysRevC.79.064324
22. C. Forssén, J.P. Vary, E. Caurier, P. Navrátil, Phys. Rev. C **77**, 024301 (2008), http://link.aps.org/doi/10.1103/PhysRevC.77.024301
23. P. Maris, J.P. Vary, A.M. Shirokov, Phys. Rev. C **79**, 014308 (2009), http://link.aps.org/doi/10.1103/PhysRevC.79.014308
24. Y. Tang, M. LeMere, D. Thompsom, Phys. Rep. **47**, 167 (1978), ISSN 0370-1573, http://www.sciencedirect.com/science/article/pii/0370157378901758
25. K. Langanke, H. Friedrich, Adv. Nucl. Phys. **17**, 223–363 (1986)
26. http://www.int.washington.edu/
27. http://unedf.org/
28. P. Sternberg, E.G. Ng, C. Yang, P. Maris, J.P. Vary, M. Sosonkina, H.V. Le, in *Proceedings of the 2008 ACM/IEEE Conference on Supercomputing SC*, 08, (IEEE Press, Piscataway, 2008) pp. 15:1–15:12. ISBN 978-1-4244-2835-9. http://dl.acm.org/citation.cfm?id=1413370.1413386
29. P. Maris, M. Sosonkina, J.P. Vary, E. Ng, C. Yang, Procedia Comput. Sci. **1**, 97, (2010), ISSN 1877-0509, http://www.sciencedirect.com/science/article/pii/S187705091000013X
30. http://science.energy.gov/ascr/research/scidac/
31. http://science.energy.gov/ascr/facilities/incite/about-incite/
32. http://www.frib.msu.edu/

Chapter 2
The No Core Shell Model

2.1 Introduction

As has been discussed in the introduction, the No-Core Shell Model (NCSM), is one of the recently developed *ab-initio* many-body techniques, for solving bound state properties of light nuclei ($A \leq 20$) [1–3]. The NCSM is different from past shell model calculations, since we allow for all A nucleons to be active in the model-space. The traditional shell model assumes that only the valence nucleons are active in the model-space. Another key difference to the standard shell model, is the use of realistic NN and NNN interactions [4]. In the past, effective interactions were derived for the valence space nucleons, an example of which are the USD effective interactions [5]. In the NCSM, we try to move away from such techniques, in order to stay true to our goal of calculating nuclear-structure directly from realistic interactions that are based on QCD.

In Sect. 2.2, I will discuss the nuclear Hamiltonian employed. I will also briefly describe how we solve it in the NCSM. The Hamiltonian requires knowledge of the nuclear interaction. I discuss some of the common types of nuclear interactions in use (see Sect. 2.2.1) and how they are modified through various renormalization procedures (Sect. 2.2.2) for the NCSM. The NCSM basis, as well as the various consequences of using either relative or single-particle coordinates is discussed in Sect. 2.3. In Sect. 2.4 I show some sample NCSM calculations and discuss how the NCSM converges with respect to the size of the basis. I then illustrate what the current limits of the NCSM are (see Sect. 2.5). The extensions of the NCSM beyond these limits will be the bulk of the thesis, which I will discuss extensively in Chaps. 3, 4 and 5. In Sect. 2.5.1 I make a note on current NCSM computer codes in use today, as well as commenting on various implementations of many-body physics in these codes.

M. K. G. Kruse, *Extensions to the No-Core Shell Model*,
Springer Theses, DOI: 10.1007/978-3-319-01393-0_2,
© Springer International Publishing Switzerland 2013

2.2 The Nuclear Hamiltonian

We begin with the translationally invariant nuclear Hamiltonian,

$$H_A = \frac{1}{A}\sum_{i<j}^{A}\frac{(\vec{p}_i - \vec{p}_j)^2}{2m} + \sum_{i<j}^{A} V_{\mathrm{NN},ij} + \sum_{i<j<k}^{A} V_{\mathrm{NNN},ijk} \tag{2.1}$$

in which, m is the mass of the nucleon and A is the number of nucleons present (i.e., the mass number). The first term represents the relative kinetic energy, the second term represents the nucleon-nucleon (NN) interaction, and the third term represents the three nucleon (NNN) interaction. In what follows, we will only work with NN forces, so we disregard the last term in Eq. (2.1), even though it is known to be important for agreement with experiment. Since the bulk of this thesis deals with extending the capabilities of the NCSM, we are not too concerned with missing NNN contributions, since we are not attempting to agree with experimental data. Many of the extensions that I will discuss can be extended in a fairly straightforward manner so as to include the NNN forces.

The NCSM employs the three-dimensional isotropic harmonic oscillator (HO) functions, which are particularly convenient for the center-of-mass separation, as I will discuss in Sect. 2.3. In the absence of any interactions among the nucleons, the A nucleons will fill the harmonic oscillator levels in accordance with the Pauli exclusion principle. We will refer to this situation as the unperturbed ground state (gs) configuration. The number of oscillator quanta that represent this non-interacting picture is given by $N_{\mathrm{min}} = \sum_i^A (2n_i + l_i)$. For example, in the case of ^{7}Li, $N_{\mathrm{min}} = 3$, since 3 valence nucleons occupy the $l = 1$ 0p-shell. Recall that the other 4 nucleons occupy (and fill) the 0s-shell. The computational model-space is characterized by a parameter N_{max}, which defines how many oscillator quanta of energy can be shared among the A nucleons, *above* the unperturbed ground-state (gs) energy. For example, an $N_{\mathrm{max}} = 4$ ^{7}Li calculation would allow at most $N_{\mathrm{min}} + N_{\mathrm{max}} = 7$ quanta of energy to be available to the 7 nucleons. The two-body Hamiltonian could thus promote a valence neutron in the $N = 1$ shell, to the $N = 5$ shell. Alternatively, two valence neutrons could be simultaneously promoted from the $N = 1$ shell to the $N = 3$ shell. All possible combinations that respect the symmetries of the nuclear Hamiltonian define the total number of states present in the basis. Note that for example, in our $N_{\mathrm{max}} = 4$ ^{7}Li example, we cannot promote one valence neutron to the $N = 2$ shell, and another valence neutron to the $N = 3$ shell. In that particular example, we are starting with a negative parity state, and by promotion of the two neutrons, ending up in a positive parity state. Since parity is conserved in nuclear interactions, we are forbidden to allow such configurations.

2.2.1 Nuclear Interactions

A wide variety of high-quality interactions exist, which reproduce scattering data, i.e., phase shifts, up to roughly 300 MeV. These interactions, when fitted to the Nijmegen scattering data [6], reproduce the data in various angular momentum channels (e.g., the $^1S_0$ or $^3S_1$ channel), with a $\chi^2/\text{dof} \sim 1$. I will briefly list some of interactions in use today.

In 1935 Yukawa hypothesized that the nuclear force between nucleons is mediated by the (then undiscovered) pions [7]. In the decades that followed Yukawa's initial idea, generations of meson exchange interactions were constructed, in which not only π's, but also other mesons such as the ρ and ω were included as nuclear-force mediators. One of the most successful NN meson exchange interactions that was constructed is the charge- and momentum-dependent Bonn potential (e.g., CD-Bonn) [8]. Another popular NN interaction is the most recent Argonne potential, Av_{18} [9], which has a number of terms such as spin-spin, spin-orbit and tensor contributions included. The various terms are fitted by experimental data. A NN interaction that attempts to mimic the missing 3N forces is, for example, the INOY potential (Inside Non-local, Outside Yukawa) [10, 11], in which a density dependent two-nucleon force is created. Density-dependent interactions change their relative strength as the number of nucleons are increased. Unfortunately, none of these potentials has a direct connection to QCD, nor is there any clear hierarchy of which terms are the dominant ones. These potentials cannot be used directly in Eq. (2.1), since they generate strong short-range correlations. This means that a perturbative treatment, or even a diagonalizaton technique, converges very slowly in the number of terms calculated or with the size of the basis. These are, thus, often referred to as hard(-core) potentials, since they invoke the schematic picture of the strong repulsive potential barrier found in the NN forces at small distances.

2.2.1.1 The Chiral Interaction

The fact that QCD is non-perturbative at energy scales relevant for nuclear-structure, puts nuclear physicists in a tough spot. Ideally, we would like to perform calculations of light nuclei, that have a direct connection to QCD. This connection was provided to the structure community by means of Effective Field Theory (EFT) [12–16], which was applied to nuclear systems by pioneers such as Ordóñez, Ray, and van Kolck [17–19]. In an effective field theory, one uses the relevant degrees of freedom appropriate for the momentum scale at hand. In the case of nuclear-structure, the connection to QCD is offered by the chiral EFT Lagrangian, in which the long- and medium-range interactions in a nucleus are described by nucleons interacting through pion exchange. The short range physics is integrated out and is expressed as contact terms, which requires the introduction of low-energy constants.

The development of chiral effective field theory provided the connection to QCD that structure physicists were desperately lacking. It also provided a clear and consice

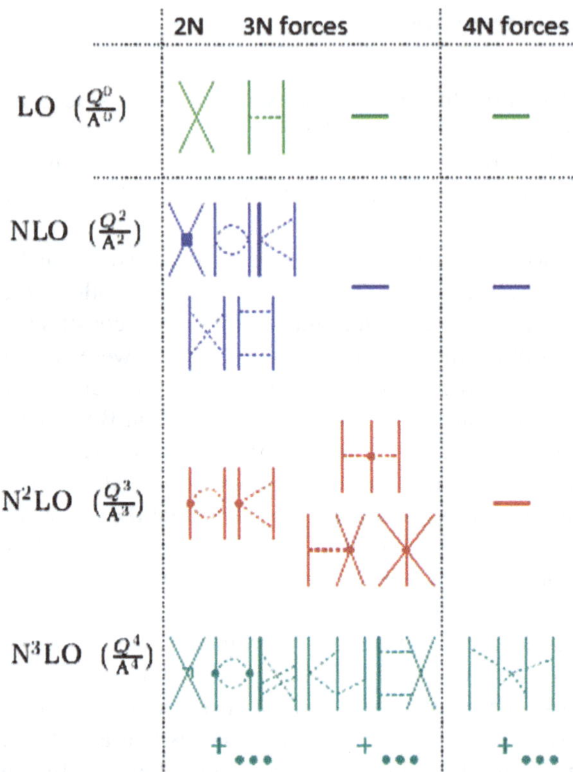

Fig. 2.1 The various perturbation orders of chiral EFT. *Solid lines* indicate nucleons interacting via pions (*dashed lines*). At order N2LO, a "three-body" force naturally arises. The *solid dots* located at the vertices, for instance at N2LO, describe low-energy constants (LEC's), that need to be determined from experimental data. The figure is adapted from [21]

picture of the nuclear force, organized in a hierarchy of terms, arranged according to a power-counting scheme.[1] This is possible, since chiral EFT is pertubative at these momentum scales, unlike QCD. A pictorial description of the various terms arising in EFT, can be seen in Fig. 2.1. Each order in the perturbation series is organized according to a power counting scheme, the details of which are not too important, except that each higher-order term is smaller than the previous terms. What is however important, is the determination of the low-energy constants (LEC) that arise in chiral EFT. An example of an LEC can be seen in Fig. 2.1 for the N2LO order, indicated by the solid dot on a vertical line. These constants in some sense are a consequence of our ignorance of the full QCD theory, and cannot be determined theoretically. They can however be determined from experiments or be constrained by nuclear-structure calculations [20].

[1] The Weinberg power counting scheme is known to be inconsistent. As there is no other viable EFT alternative for the many-body community presently, it is often used in calculations without regard to the issues of renormalization-group invariance.

For nuclear-structure calculations, the EFT interaction of choice is the 'Entem and Machleidt' interaction, in which NN terms are kept up to order N3LO [22] and NNN terms are kept up to order N2LO [23].

2.2.1.2 Softened Interactions

Recently, a new class of interactions has emerged, commonly known as soft-potentials. As the name suggests, these interactions have the repulsive hard-core of the interaction softened in some manner. The interactions themselves are not new, in the sense that they are not constructed from any underlying *physical* theory. Instead, the initial interactions, such as the chiral N3LO NN interaction, is transformed by a series of unitary transformations, such as the $v_{\text{low-k}}$ method (see [24] and references therein), or more recently, the similarity renormalization group (SRG) method [25–28]. These methods decouple the high-momentum components of the bare interaction from the low-momentum components. In doing so, the effect of the repulsive hard-core to generate strong short-range correlations is lessened. Thus, the softened interactions can be used directly in Eq. (2.1) without any further modifications to the Hamiltonian itself. The use of SRG in nuclear systems has been demonstrated in [29–32]. In the case of bare interactions, one usually has to create an effective interaction for the NCSM space; a procedure known as the Okubo-Lee-Suzuki method [33, 34], which I will describe in Sect. 2.2.2. The most dramatic difference in using the softened interactions, when compared to the initial bare interactions, is that they speed up the rate of convergence of the many-body calculations, as the model-space size is increased. This is in part due to the decoupling of high- and low-momentum components in the softened interactions. In this regard, the bare chiral N3LO NN interaction is 'softer' than other bare interactions, allowing for the possibility to use it's bare form in Eq. (2.1) directly. However, with the current computer resources at hand, convergence of the NCSM with the bare chiral interaction has only been demonstrated for the Triton and ^{4}He [32, 35]. For heavier nuclei, the chiral interaction must either be renormalized by the Lee-Suzuki procedure, or be softened by means of the SRG procedure (or by a similar procedure) [36].

2.2.2 *Effective Interactions*

Before softened interactions became available, it was necessary to derive an effective NN interaction, specific for a given $N_{\max}$ space, derived from the original NN interaction. From a practical view this is necessary, in order to have the calculations converge in a tractable basis space. To speed up the convergence of the calculations, we modify Eq. (2.1), by adding to it the harmonic oscillator center-of-mass term, $H_{CM} = \frac{A\vec{P}^2}{2m} + \frac{1}{2}mA\Omega^2\vec{R}^2$, in which $\vec{R} = \frac{1}{A}\sum_i^A \vec{r}_i$, is the center-of-mass coordinate. A similar expression exists for the center of momentum term, $\vec{P} = \frac{1}{A}\sum_i^A \vec{p}_i$.

$$H_A^{\Omega} = H_A + H_{CM} = \sum_{i=1}^{A} h_i + \sum_{i<j}^{A} V_{ij}^{\Omega,A} \tag{2.2}$$

$$= \sum_{i=1}^{A} \left[\frac{\vec{p}_i^{\,2}}{2m} + \frac{1}{2} m\Omega^2 \vec{r}_i^{\,2} \right] + \sum_{i<j}^{A} \left[V_{\mathrm{NN},ij} - \frac{m\Omega^2}{2A} (\vec{r}_i - \vec{r}_j)^2 \right] \tag{2.3}$$

Note that after the addition of the center-of-mass term, the Hamiltonian is dependent on A, as well as the chosen HO frequency Ω. The chosen value of Ω does affect the calculated gs energy, although, in a variational way. The optimal choice of Ω, for a given nucleus and specified model-space, is a value that minimizes the gs energy. In practice, a range of possible $\hbar\Omega$ values are feasible, since the dependence near the variational minimum is approximately constant. For most nucleon-nucleon interactions, one cannot use Eq. (2.3) directly for $A > 3$, since one would need an extremely large model-space, i.e, a large value for $N_{\max}$ must be used.

I will now describe the method we prefer to use to create an effective interaction from the starting A-body Hamiltonian, given in Eq. (2.3). The full details of this approach as well as the relation to the Okubo-Lee-Suzuki transformation are given in [37]. Although we cannot solve Eq. (2.3) for a general A-body system, we can solve it for the $A = 2$, and also for $A = 3$. We will use this ability as our starting point for creating an effective interaction. We begin by solving the Hamiltonian in the two-body case, $A = 2$, by employing a few hundred oscillator (HO) shells, in which typically $N_{\max} = 350$–450. For clarity, Eq. (2.3) assumes the simple form,

$$H_{A,a=2}^{\Omega} = h_1 + h_2 + V_{ij}^{\Omega,A}. \tag{2.4}$$

Equation (2.4) represents an $a = 2$ cluster approximation of the full A-body Hamiltonian. Note, however, that the two-body cluster approximation still retains information of the A-body system, through the third term on the right. By diagonalizing Eq. (2.4), we are able to solve for the eigenenergies of the $a = 2$ cluster. One can also think of the diagonalization as performing a unitary transformation on the Hamiltonian., in which U_2 are the matrices that contain the eigenvectors of Eq. (2.4).

$$E_{A,a=2}^{\Omega} = U_2 H_{A,2}^{\Omega} U_2^{\dagger} \tag{2.5}$$

The unitary matrices can be split into four blocks, each representing a different vector space.

$$U_2 = \begin{pmatrix} U_{2,P} & U_{2,PQ} \\ U_{2,QP} & U_{2,Q} \end{pmatrix} \tag{2.6}$$

The block-square $d_P \times d_P$ $U_{2,P}$ matrix represents the P-space, or the basis space ($N_{\max}$), in which the effective interaction is to be employed, whereas the Q space refers to all the eigenstates we are excluding in the effective interaction. Typically, the P-space is characterized by an $N_{\max A}$ value, which is much smaller than the

$N_{\max 2} \sim 400$ value that was used to solve the $A = 2$ case. Here, we have made a distinction between the two parameters $N_{\max A}$ and $N_{\max 2}$. The former refers explicitly to the space in which the A-body calculation will be performed, whereas the latter refers to the space in which the $A = 2$ system is solved. However, in many references, one will simply find a value stated for $N_{\max}$, without explicit distinction between the general A- or two-body case. It is usually clear from the context of the situation, which particular $N_{\max}$ definition is implied.

The U_2 matrices diagonalize the two-body Hamiltonian, and, thus, bring the Hamiltonian into diagonal form.

$$E^{\Omega}_{A,2} = \begin{pmatrix} E_{A,2,P} & 0 \\ 0 & E_{A,2,Q} \end{pmatrix} \tag{2.7}$$

We are now in a position to calculate the effective Hamiltonian, in which $N_{\max}$ now refers to the model-space of the A-body calculation.

$$H^{N_{\max},\Omega}_{A,2} = \frac{U^{\dagger}_{2,P}}{\sqrt{U^{\dagger}_{2,P} U_{2,P}}} E^{\Omega}_{A,2,P} \frac{U_{2,P}}{\sqrt{U^{\dagger}_{2,P} U_{2,P}}} \tag{2.8}$$

The approach is similar to the methods used previously in the NCSM, as shown in Sect. 2.3.1 in [1]. Equation (2.8) is also particularly convenient, as it avoids storing a large number of the ω-operator matrix elements that are required in [1].

2.3 Basis Functions in the NCSM

I will now extensively discuss the basis of the NCSM. We choose to work with a HO basis, since it is possible to separate the spurious center-of-mass states from the intrinsic states exactly, provided we work in the $N_{\max}$ truncation. Furthermore, the HO basis has well-known analytical properties that are particularly convenient for our purpose. For example, HO basis states expressed in relative coordinates are easily transformed into single-particle coordinates by means of the Talmi-Brody-Moshinksy orthogonal transformation [38]. A single-nucleon HO wavefunction, in which we temporarily neglect the spin of the nucleon, can be written as,

$$\phi_{nlm}(\vec{r}; b) = R_{nl}(r; b) Y_{lm}(\hat{r}), \tag{2.9}$$

in which $R_{nl}(r; b)$ and $Y_{lm}(\hat{r})$ correspond to the radial wavefuntion and the corresponding spherical harmonic, respectively. The radial wavefunctions depend on b and the HO parameter, which is related to the HO frequency Ω by $b = \sqrt{\frac{\hbar}{m\Omega}}$, in which m is the average nucleon mass. The oscillator parameter is seen to be a length-scale, set by the harmonic oscillator energy spacing ($\hbar\Omega$); for small

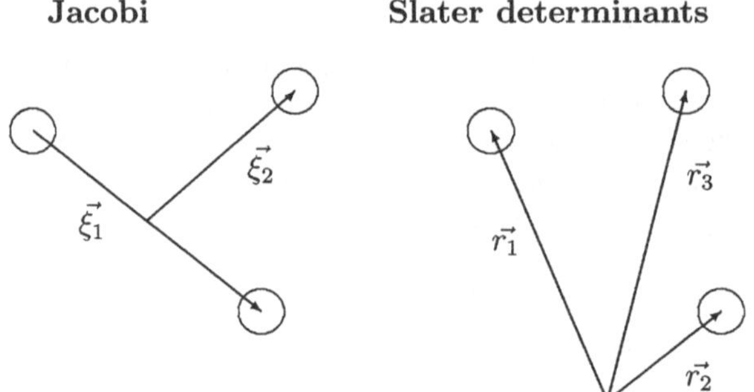

Fig. 2.2 The figure shows the difference between relative (*Jacobi*) and single-particle (*Slater*) coordinates. Note that Jacobi coordinates are defined relative to the center-of-mass of a sub-cluster, whereas the single-particle coordinates are defined from a common point

values of Ω, the oscillator well is very broad, whereas for large values of Ω, the oscillator well is very narrow. The optimal choice of $\hbar\Omega$ will be discussed in Sect. 2.4.

In order to form the basis for the A-nucleon system, one of two approaches can be pursued. The first choice is to express the many-body basis in terms of relative coordinates, as is done in the Jacobi basis [39]. This form is the natural starting point, as nuclear interactions are expressed in terms of relative momenta (or position). However, the antisymmetrization of Jacobi coordinates is very cumbersome for $A > 4$. Thus, for heavier systems, single-particle coordinates are preferred, leading to a basis consisting of Slater determinants. A schematic view of the two types of basis sets is shown in Fig. 2.2.

2.3.1 Jacobi Basis

The Jacobi basis is based on relative coordinates, $\vec{\xi_i}$, in which i labels the i-th coordinate in the basis. The first coordinate, $\vec{\xi_0}$, is taken to be the center-of-mass of the A-nucleon system. The second coordinate, $\vec{\xi_1}$, is taken to be the center-of-mass between nucleon 1 and nucleon 2. Each $\vec{\xi_i}$ thereafter is defined as the position of the $i+1$-th nucleon with respect to the center-of-mass of the first i nucleons (refer to the left diagram in Fig. 2.2).

$$\vec{\xi_0} = \sqrt{\frac{1}{A}}\left[\vec{r_1} + \vec{r_2} + \cdots + \vec{r_A}\right] \tag{2.10}$$

$$\vec{\xi_1} = \sqrt{\frac{1}{2}}\left[\vec{r_1} - \vec{r_2}\right] \tag{2.11}$$

$$\vec{\xi_2} = \sqrt{\frac{2}{3}}\left[\frac{1}{2}(\vec{r_1} + \vec{r_2}) - \vec{r_3}\right] \tag{2.12}$$

$$\vec{\xi_i} = \sqrt{\frac{A-1}{A}}\left[\frac{1}{A-1}(\vec{r_1} + \vec{r_2} + \cdots + \vec{r_{A-1}}) - \vec{r_A})\right] \tag{2.13}$$

The various numerical prefactors, as show in Eq. (2.10), are required to have an orthonormal transformation between single-particle and relative coordinates. By inverting the above expressions, it is possible to express single-particle coordinates ($\vec{r_i}$) in terms of the Jacobi coordinates ($\vec{\xi_i}$).

At this stage we have only specified the relative coordinates. We must still construct the antisymmetric basis, as I will now outline. The antisymmetrization is most easily demonstrated in the case of three nucleons ($A = 3$). Detailed discussions can be found in [2, 39, 40].One begins with the following basis state,

$$|(nlsjt; \mathcal{NLJ})JT\rangle \tag{2.14}$$

In Eq. (2.14), n and l correspond to the HO quantum numbers corresponding to the Jacobi coordinate $\vec{\xi_1}$. Note that two nucleons are involved in the definition of $\vec{\xi_1}$, thus the additional quantum numbers s, j, t correspond to the relative two-nucleon channel quantum numbers, spin, angular momentum and isospin, respectively. For example, the spin of the two-nucleon channel is defined as $s = s_1 + s_2$, whereas $j = l + s$ and $t = t_1 + t_2$. The relative two-nucleon channel is, by construction, antisymmetric with respect to particle exchange, and is mathematically expressed by the requirement that $(-1)^{s+l+t} = -1$. The quantum numbers $\mathcal{NLT}$ correspond to the quantum numbers of the third nucleon, associated with the relative coordinate $\vec{\xi_2}$. The total angular momentum and total isospin of the three-nucleon system is given by $J = j+\mathcal{J}$ and $T = t+\mathcal{T}$. Note that although the state $|(nlsjt; \mathcal{NLJ})JT\rangle$ is explicitly antisymmetric with respect to nucleon 1 and nucleon 2, it is *not* antisymmetric with respect to nucleons 1 and 3 or nucleons 2 and 3 being exchanged.

In order to construct an antisymmetric basis, we need to make use of an antisymmetrizer,

$$\mathcal{X} = \frac{1}{3}(1 + \mathcal{T}^{(-)} + \mathcal{T}^{(+)}). \tag{2.15}$$

The operators $\mathcal{T}^{(-)}$ and $\mathcal{T}^{(+)}$ are anti-cyclic and cyclic permutation operators, respectively. In order to create the antisymmetric basis, one needs to diagonalize the operator $\mathcal{X}$ in the $|(nlsjt; \mathcal{NLJ})JT\rangle$ basis. The resulting eigenvalues will span two subspaces; an eigenvalue of 1 corresponds to a physically antisymmetric state, whereas an eigenvalue of 0 corresponds to a spurious (non-antisymmetric) state [41]. In order to diagonalize $\mathcal{X}$, one needs to determine the matrix elements of the operator. Since the $|(nlsjt; \mathcal{NLJ})JT\rangle$ basis is already antisymmetric with respect to nucleons 1 and 2, we can determine the matrix elements of $\mathcal{X}$ from the action of the permutation operator P_{23}, which interchanges nucleon 2 and nucleon 3.

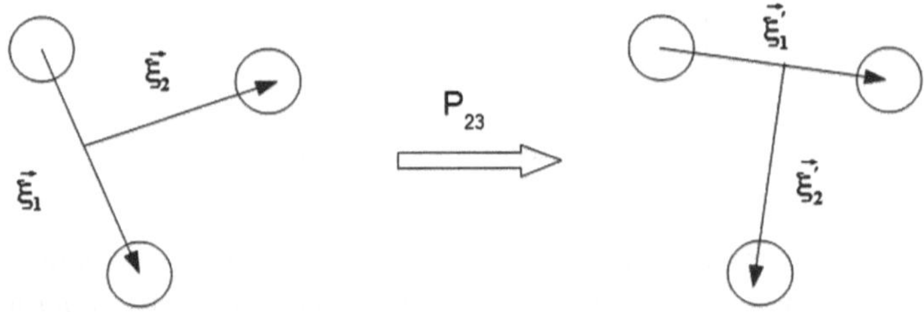

Fig. 2.3 The figure shows the action of the permutation operator P_{23} on the basis state $|(nlsjt(\vec{\xi}_1); \mathcal{NLJ}(\vec{\xi}_2)JT\rangle$, in which nucleon 2 and 3 are exchanged. The original state, $|(nlsjt(\vec{\xi}_1); \mathcal{NLJ}(\vec{\xi}_2)JT\rangle$ (*left*) is connected to the new state $|(nlsjt(\vec{\xi}'_1); \mathcal{NLJ}(\vec{\xi}'_2)JT\rangle$ (*right*) by means of an orthogonal transformation

$$\langle \mathcal{X} \rangle = \frac{1}{3}\langle (1 - 2P_{23}) \rangle \tag{2.16}$$

The action of P_{23} on the basis state $|(nlsjt(\vec{\xi}_1); \mathcal{NLJ}(\vec{\xi}_2)JT\rangle$ interchanges nucleons 2 and 3. This leads to a new basis state, expressed in primed coordinates as $|(nlsjt(\vec{\xi}'_1); \mathcal{NLJ}(\vec{\xi}'_2)JT\rangle$. Note that in the primed coordinates, $\vec{\xi}'_1$ corresponds to the relative two-nucleon channel consisting now of nucleons 1 and 3. The action of P_{23} on $|(nlsjt(\vec{\xi}_1); \mathcal{NLJ}(\vec{\xi}_2)JT\rangle$ can be seen in Fig. 2.3.

The primed Jacobi coordinates can be expressed in terms of the unprimed coordinates, since the two are related by an orthogonal transformation. Thus, one can express the state $|(nlsjt(\vec{\xi}'_1); \mathcal{NLJ}(\vec{\xi}'_2)JT\rangle$ in terms of the unprimed state, $|(nlsjt(\vec{\xi}_1); \mathcal{NLJ}(\vec{\xi}_2)JT\rangle$. The expansion coefficients that enter in the orthogonal transformation are just the generalized harmonic-oscillator brackets with the mass ratio d [42], in which d is determined from the orthogonal transformation (see Eq. 2.20). Using the preceding information, it is relatively straightforward to calculate the matrix elements of the P_{23} operator, expressed in the unprimed coordinates as follows,

$$\begin{aligned}
&\langle (n_1 l_1 s_1 j_1 t_1; \mathcal{N}_1 \mathcal{L}_1 \mathcal{J}_1) JT | P_{23} | (n_2 l_2 s_2 j_2 t_2; \mathcal{N}_2 \mathcal{L}_2 \mathcal{J}_2) JT \rangle \\
&\quad = \delta_{N_1,N_2} \hat{t}_1 \hat{t}_2 \begin{Bmatrix} \frac{1}{2} & \frac{1}{2} & t_1 \\ \frac{1}{2} & T & t_2 \end{Bmatrix} \\
&\quad \times \sum_{LS} \hat{L}^2 \hat{S}^2 \hat{j}_1 \hat{j}_2 \hat{\mathcal{J}}_1 \hat{\mathcal{J}}_2 \hat{s}_1 \hat{s}_2 (-1)^L \begin{Bmatrix} l_1 & s_1 & j_1 \\ \mathcal{L}_1 & \frac{1}{2} & \mathcal{J}_1 \\ L & S & J \end{Bmatrix} \begin{Bmatrix} l_2 & s_2 & j_2 \\ \mathcal{L}_2 & \frac{1}{2} & \mathcal{J}_2 \\ L & S & J \end{Bmatrix} \\
&\quad \times \begin{Bmatrix} \frac{1}{2} & \frac{1}{2} & s_1 \\ \frac{1}{2} & S & s_2 \end{Bmatrix} \langle n_1 l_1 \mathcal{N}_1 \mathcal{L}_1 L | n_2 l_2 \mathcal{N}_2 \mathcal{L}_2 L \rangle_3 .
\end{aligned} \tag{2.17}$$

In Eq. (2.17), $N_i = 2n_i + l_i + 2\mathcal{N}_i + \mathcal{L}_i$ labels the total energy of the state $i = 1, 2$. The δ_{N_1,N_2} ensures that the total energy energy is conserved in the transformation.

The notation $\hat{j}$ is a short-hand for writing $\sqrt{2j+1}$ and $\langle n_1 l_1 \mathcal{N}_1 \mathcal{L}_1 L | n_2 l_2 \mathcal{N}_2 \mathcal{L}_2 L \rangle_3$ are the generalized harmonic-oscillator brackets with mass ratio $d = 3$.

Recall that we have to diagonalize $\mathcal{X}$. Once we have identified the antisymmetric states, we are able to express a fully antisymmetric state, $|NiJT\rangle$, in which the state is antisymmetric under *any* nucleon exchanges as follows,

$$|NiJT\rangle = \sum \langle nlsjt; \mathcal{N}\mathcal{L}\mathcal{J} || N_i JT \rangle |(nlsjt; \mathcal{N}\mathcal{L}\mathcal{J}) JT \rangle. \tag{2.18}$$

The additional label i is required to distinguish states that correspond to the same NJT quantum numbers. Furthermore, $N = 2n + l + 2\mathcal{N} + \mathcal{L}$. Finally, the expansion coefficients are given in terms of coefficients of fractional parentage $\langle nlsjt; \mathcal{N}\mathcal{L}\mathcal{J} || N_i JT \rangle$. For $A = 4$, similar expressions can be derived [40].

2.3.2 Slater Determinant Basis

We now turn our attention to constructing a basis based on single-particle coordinates, $\vec{r}_i$, in which $i = 1, 2, .., A$, label the single-particle coordinates of the A nucleons. The single-particle basis with which we choose to work, is once again based on HO single-particle states, $\phi_{nljm_j}(\vec{r}, \vec{\sigma}; b)\chi_{tm_t}(\vec{\tau})$. The position is specified by $\vec{r}$, whereas the intrinsic spin and isospin are denoted by the labels $\vec{\sigma}$ and $\vec{\tau}$, respectively. Recall that $\phi_{nljm_j}(\vec{r}, \vec{\sigma}; b) = (\phi_{nlm_l}(\vec{r}; b)\chi_{sm_s}(\vec{\sigma}))^{(j)}_{(m)}$, in which we denote the coupling of orbital angular momentum with the intrinsic spin as $j = l + s$. The state $\phi_{nlm_l}(\vec{r}; b)$ is given by the solution to the three-dimensional isotropic oscillator, $\phi_{nlm_l}(\vec{r}; b) = R_{nl}(r; b)Y_{lm_l}(\hat{r})$.

In practice, it is easiest to work with the single-particle state $\phi_{nljm_j}(\vec{r}, \vec{\sigma}; b)$ and to explicitly distinguish between protons and neutrons, making the label for isospin superfluous. Recall that for each j, there are $2j + 1$ single-particle states. Each of the single-particle states has a magnetic quantum number m_j associated with it, in which the possible values are $-j \leq m_j \leq j$. Allowing the A nucleons to occupy any of the many single-particle states leads to a total projection number $M = \sum_{i=1}^{A} m_{j,i}$. Typically, we restrict the possible values of M to either $M = 0$ for even A nuclei or $M = \frac{1}{2}$ for odd A nuclei. The construction is most easily understood in terms of Fig. 2.4, in which we have shown one possible $M = 0$ basis state for ^{6}Li in the $N_{\text{max}} = 0$ space. Note that we don't show the other four nucleons occupying the $0s_{\frac{1}{2}}$ shell, since at $N_{\text{max}} = 0$ those four nucleons are frozen energetically.

The number of basis states that are constructed are determined by the value of M. In the case of ^{6}Li at $N_{\text{max}} = 0$ (as shown in Fig. 2.4), there are 10 possible basis states. This particular scheme is known as the Glasgow m-scheme [43], in which basis states are solely constructed on the total M. The m-scheme is particularly efficient as it requires no tedious angular momentum coupling,which would typically involve coefficients of fractional parentage. All the antisymmetrization is taken care by the single-particle nature of the occupation scheme, which also makes the use of

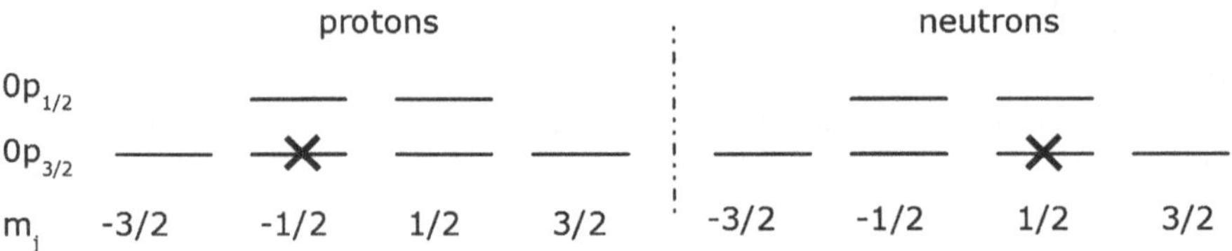

Fig. 2.4 The figure shows one of 10 possible basis states (for which $M = 0$) that can be formed for ^{6}Li at $N_{\max} = 0$. The *cross* indicates the particular single-particle state that is occupied. Note that in order to form an $M = 0$ state, we need to place the proton and neutron into single-particle states such that the individual m_j sum up to zero

second-quantization methods particularly appealing. In this regard, single-particle state occupations are represented in a bit fashion (i.e., '1' represents an occupied state, whereas a '0' represents an occupied state). The computational price we pay is that our Hamiltonian matrix, expressed in terms of the m-scheme basis is very large, and that J is no longer a good quantum number. Nuclear states are characterized by $J^{\pi}T$, in which J^{π} corresponds to the total angular momentum and parity of the state and where T labels the total isospin of the state. Good J and T quantum numbers are recovered in the m-scheme by projecting the final wavefunction onto good J and T.

2.3.3 The Center-of-Mass Issue

The use of single-particle coordinates breaks the translational invariance of the nuclear Hamiltonian (for extensive discussion see [44] and the references therein). In the definition of the coordinates, $\vec{r}_i$, we have implicitly defined a point in space, from which all position (or momenta) originate. The consequence of this choice is that our calculated wavefunctions, $\Psi(\vec{x})$, contain spurious center-of-mass states. These states describe the motion of the center-of-mass of the nucleus, and have nothing to do with the interesting intrinsic states that we are attempting to calculate. In other words, we would like to decompose the total wavefunction as a product of intrinsic ($\phi_{\text{int.}}(\vec{x})$) and center-of-mass states (CM) as follows, $\Psi(\vec{x}) = \phi_{\text{int.}}(\vec{x}) \otimes \phi_{\text{CM}}(\vec{X})$. The question is, how does one separate spurious center-of-mass states from the intrinsic states? In the case of the NCSM, one can separate the two types of states exactly, provided two things are taken into consideration.

The first consideration is that the HO basis be truncated on an energy-quanta level, i.e., the $N_{\max}$ truncation is used. If a single-particle truncation is used, as is done in Hartree-Fock, then one mixes the center-of-mass state with the intrinsic states. Note that we explicitly stated that HO basis functions be used; any other basis set, such as the Woods-Saxon basis, cannot decompose the center-of-mass states from the intrinsic states. This feature of the HO basis is really a consequence of the transformation properties of the basis, as I will now describe. The transformation of HO wavefunctions expressed in single-particle coordinates $\vec{r}_i$ to relative and center-of-mass HO wavefunctions expressed in coordinates $(\vec{r}, \vec{R})$, is given by the following expression,

$$\left(\phi_{n_1l_1}(\vec{r_1})\phi_{n_2l_2}(\vec{r_2})\right)^{(\lambda)}_{(\mu)} = \sum_{nl,NL} \langle nl, NL, \lambda | n_1l_1, n_2l_2, \lambda\rangle_d \left(\phi_{nl}(\vec{r})\phi_{NL}(\vec{R})\right)^{(\lambda)}_{(\mu)}, \tag{2.19}$$

provided the single-particle coordinates $\vec{r}_i$ are related to the relative $(\vec{r})$ and center-of-mass coordinate $(\vec{R})$ by

$$\begin{aligned}\vec{r} &= \left(\frac{d}{1+d}\right)^{\frac{1}{2}} \vec{r}_1 - \left(\frac{1}{1+d}\right)^{-\frac{1}{2}} \vec{r}_2 \\ \vec{R} &= \left(\frac{1}{1+d}\right)^{-\frac{1}{2}} \vec{r}_1 + \left(\frac{d}{1+d}\right)^{\frac{1}{2}} \vec{r}_2.\end{aligned} \tag{2.20}$$

In Eq. (2.19), we have written the product of two single-particle HO wavefunctions, when coupled to angular momentum λ (projection μ), as an expansion over relative and center-of-mass HO wavefunctions, in which the expansion coefficients are the generalized HO brackets with mass ratio d. Note that the transformation conserves the total energy, by enforcing that $2n_1 + l_1 + 2n_2 + l_2 = 2n + l + 2N + L$, conserves the angular momentum (λ), and also conserves parity $((-1)^{l_1+l_2} = (-1)^{l+L})$. It is *this* transformation, as shown in Eq. (2.19), that guarantees the separation of center-of-mass from the intrinsic states is exact, but only if the $N_{\max}$ truncation is used (due to the conservation of energy requirement).

I will now resume the discussion concerning the decomposition into physical and center-of-mass states. The A-nucleon physical state of interest,

$$\phi_{\text{int.}}(\vec{x}) = \langle \vec{r_1} \ldots \vec{r_A} \sigma_1 \ldots \sigma_A \tau_1 \ldots \tau_A | A\alpha J M T M_T\rangle_{SD},$$

is one in which the center-of-mass is in the lowest energy state, i.e., it is in the $0\hbar\Omega$ configuration. The label α is an additional quantum label used to distinguish between states that have the same JT assignment.

$$\begin{aligned}&\langle \vec{r_1} \ldots \vec{r_A} \sigma_1 \ldots \sigma_A \tau_1 \ldots \tau_A | A\lambda J M T M_T\rangle_{SD} \\ &\quad = \langle \vec{\xi_1} \ldots \vec{\xi}_{A-1} \sigma_1 \ldots \sigma_A \tau_1 \ldots \tau_A | A\lambda J M T M_T\rangle \times \phi_{000}(\vec{\xi_0}; b)\end{aligned} \tag{2.21}$$

In order to guarantee that the center-of-mass state is in the $0\hbar\Omega$ configuration, as shown in Eq. (2.21), we need to make use of the Gloeckner-Lawson projection method [45]. This is the second requirement that the center-of-mass states do not contaminate the intrinsic spectrum. The Lawson projection method is implemented by adding to the intrinsic Hamiltonian a HO center-of-mass term, H_{CM}, multiplied by a Lagrange multiplier, β.

Table 2.1 11 $J = 1^+$ states calculated for ^{6}Li in $(N_{\max}, \hbar\Omega) = (4, 20\,\text{MeV})$, in which the Lagrange multiplier is set to $\beta = 0.5$

Deg.	E [MeV]	Comment
1	−26.658	–
1	−24.491	–
1	−17.576	–
1	−13.560	–
3	−11.784	CM state
1	−11.299	–
3	−9.452	CM state

We use the effective CD-Bonn 2000 interaction. The first column indicates the degeneracy of the calculated state, whereas the third column indicates if a state is a spurious center-of-mass state. Note that the spurious states are easy to identify as they are usually degenerate. At $\beta = 10$, the spurious states no longer appear in the calculated spectrum

$$H = H_A + \beta\left(H_{CM} - \frac{3}{2}\hbar\Omega\right) \tag{2.22}$$

Note the center-of-mass state in the $0\hbar\Omega$ configuration will have no effect on the intrinsic energy spectrum, since the term $(H_{CM} - \frac{3}{2}\hbar\Omega)\phi_{000}(\vec{\xi}_0; b) = 0$. The center-of-mass states not in the $0\hbar\Omega$ configuration will be shifted upwards in energy, relative to the gs, on the order of $\beta\hbar\Omega$ MeV. Typically, we take $\beta = 10$. A good rule of thumb is to take a value of $\beta\hbar\Omega$ that is a few times larger than the excitation energy of the highest lying state one is interested in calculating. As an example of a spectrum, in which spurious center-of-mass states are present, we refer to Table 2.1, in which we have calculated 11 $J = 1^+$ states for ^{6}Li, using the effective CD-Bonn 2000 interaction, in which we have included the Coloumb interaction. The calculation is done for $(N_{\max}, \hbar\Omega) = (4, 20\,\text{MeV})$, except that we have only set $\beta = 0.5$. Thus, we expect at about 10 MeV in excitation energy, spurious center-of-mass states to appear.

It is worth pointing out that this issue with the center-of-mass state is an inconvenience for nuclear-structure calculations. In the case of atomic-structure calculations, one does not need to take this issue into account, since in that case the atomic nucleus itself defines the center-of-mass of the atomic system. Furthermore, a typical atomic structure Hamiltonian (consider the Born-Oppenheimer approximation), has no translational invariance to begin with, since the electronic coordinates are defined with respect to the location of the atomic nucleus.

2.4 Convergence Properties

I will now discuss some of the general features of the NCSM, in particular, how a realistic calculation is performed in practice. As an example case, we will calculate the gs of the ^{6}Li nucleus. This nucleus captures many of the features of the NCSM,

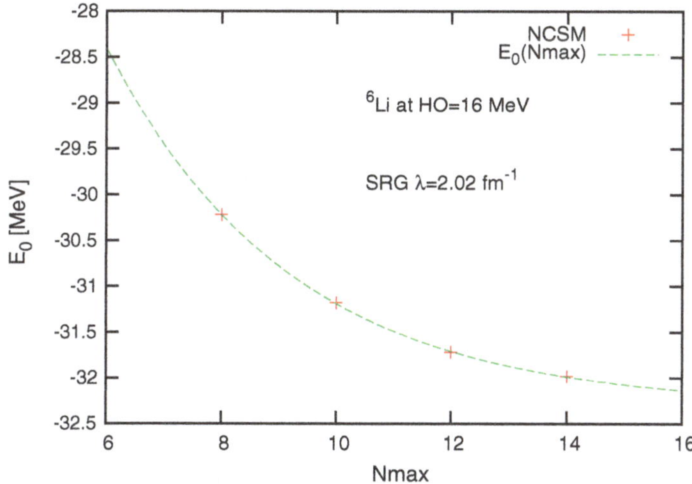

Fig. 2.5 The gs energy of ^{6}Li calculated as a function of N_{max}. We have used the fixed HO energy of $\hbar\Omega = 16$ MeV. The calculations are variational in N_{max}, if we use the bare (or SRG evolved) interaction, as done here. The calculation has not yet converged at $N_{max} = 14$, so we perform an extrapolation to $N_{max} = \infty$, which is shown by the *dashed line*. The extrapolated result is $E_\infty = -32.304$ MeV

as well as exposing us to the challenges of large-scale many-body calculations. Furthermore, it is the heaviest nucleus for which we are able to obtain near-converged results, using the SRG transformed chiral NN N3LO interaction.

Before we show the results of the NCSM calculation, it is important to define exactly what we mean by *convergence*. By convergence of a NCSM calculation, we are making a statement that the ground-state energy of a particular nucleus has been calculated (or estimated) without any dependence on the NCSM parameters. As was stated in Sect. 2.2, the NCSM depends on two parameters, N_{max} and $\hbar\Omega$. N_{max} determines the size of the basis by specifying how many single-particle states are included in the calculation, whereas the $\hbar\Omega$ dependence is a result of using HO wavefunctions for the single-particle states. In other words, when we say a calculation is converged, we mean to say that the calculation is free of any dependencies on these two quantities. In practice, one tries to make N_{max} as large as possible, and then extrapolates the gs energy as a function of N_{max}, at a fixed value of $\hbar\Omega$. This procedure removes the dependence on N_{max}. However, what particular value to choose for $\hbar\Omega$ is usually not known before hand. Thus, in practice, a series of calculations are performed, employing a variety of $\hbar\Omega$ values. The optimal choice is usually taken as the value of $\hbar\Omega$ that minimizes the gs energy in the largest possible N_{max} space.

In Fig. 2.5 we show the result of a NCSM calculation, in which we have calculated the gs energy of ^{6}Li at $\hbar\Omega = 16$ MeV. Note that as we increase the size of the basis (i.e., N_{max}), the energy decreases monotonically, as is expected from a variational calculation. The variational nature of such a calculation allows us to perform

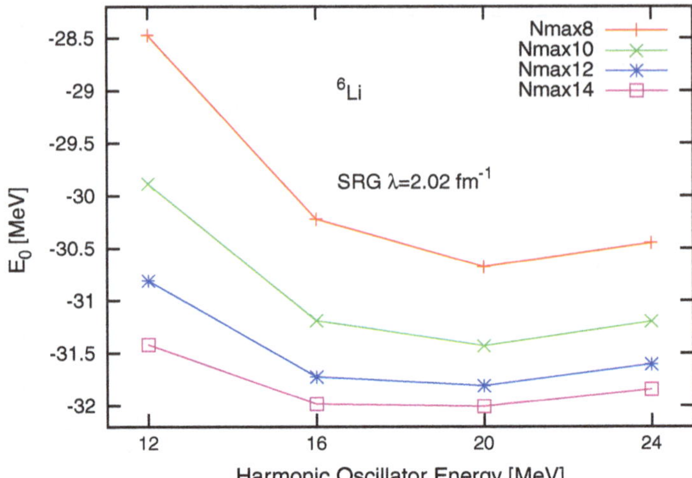

Fig. 2.6 The dependence of the gs energy on the harmonic oscillator energy $\hbar\Omega$, shown as a function of $N_{\max}$. Note that the curves are variational in $\hbar\Omega$ and $N_{\max}$. The gs energy starts to be independent of $\hbar\Omega$ as $N_{\max}$ increases for a range of $\hbar\Omega$ values, signaling that the calculations are independent of $\hbar\Omega$. The interaction is the chiral NN N3LO interaction, softened by the SRG procedure to $\lambda = 2.02\,\text{fm}^{-1}$

a suitable extrapolation to $N_{\max} = \infty$ by using an exponential decay, given as

$$E_0(N_{\max}) = a * \exp(-b * N_{\max}) + E_{\infty}. \tag{2.23}$$

At $N_{\max} = \infty$, we determine the gs energy to be E_{∞}; for the calculation shown in Fig. 2.5, $E_{\infty} = -32.304$ MeV.

However, the extrapolation shown in Fig. 2.5 might still contain some dependence on the harmonic oscillator energy $\hbar\Omega$. In Fig. 2.6, we show the gs energy of ^{6}Li as a function of $N_{\max}$ and $\hbar\Omega$. Plotting the gs energies in this fashion allows us to determine what the optimal value of $\hbar\Omega$ should be. In particular, note that as $N_{\max}$ increases, the gs energy is approximately the same for a range of $\hbar\Omega$ values. This is easily seen in the plateau that is developing at $16 \leq \hbar\Omega \leq 20$ MeV at $N_{\max} = 14$. It is often argued that once the plateau has developed, any of the possible values of $\hbar\Omega$ found in that range are acceptable values to use. Frequently, the value of $\hbar\Omega$ that is chosen is the one that corresponds to the variational minimum in the largest $N_{\max}$ space. In the case of Fig. 2.6, it can be seen that this corresponds to $\hbar\Omega \approx 20$ MeV.

Once an $\hbar\Omega$ value has been chosen, usually close to the variational minimum, one can proceed to extrapolate the gs energy, as was shown in Fig. 2.5. This is one particular strategy that is used to claim independence of the NCSM parameters in the final calculation [46, 47]. A slightly more rigorous method to absolve the calculations of NCSM parameters is the Full No Core Shell Model calculations of James Vary and Pieter Maris [48]. In that case, several extrapolations to $N_{\max} = \infty$ are performed over a range of $\hbar\Omega$ values, which allows them to determine a theoretical error of the extrapolations, which indicates the uncertainty in the extrapolation due to residual dependencies on $N_{\max}$ and $\hbar\Omega$.

2.5 Current Limits of the NCSM

The NCSM is a very powerful tool for nuclear-structure calculations. By far the greatest advantage over traditional shell model approaches is the ability to use realistic interactions, while performing calculations in which all A nucleons are active in the model-space. However, calculations such as those done in the NCSM come at a steep computational price. I will now discuss the current technological limits of the NCSM and set the stage for the main parts of this thesis: extending the capabilities of the NCSM in various ways.

The 0s shell nuclei, such as the Triton (^{3}H) and the α-particle can be calculated with extremely high precision, even when using the bare chiral (NN+NNN) interaction [39]. These nuclei pose no problem on the computational side of NCSM calculations, provided the Jacobi basis is used. However, the p-shell nuclei, for which $4 < A \leq 16$, are quite a bit more computationally intensive. There are a couple of reasons for this, which I will describe in turn.

As mentioned in the previous paragraph, for 0s-shell nuclei, we can make use of the efficient Jacobi-coordinate basis. However, once $A > 4$, the antisymmetrization of the Jacobi basis becomes extremely difficult to perform. Furthermore, even for $A = 4$, in which the anti-symmetrization is still feasible, problems arise, as $N_{\max}$ increases beyond $N_{\max} = 18$. Each basis state must be orthonormal to all others already calculated; at large $N_{\max}$ values small numerical inaccuracies are introduced making the orthonormalization only approximate. Returning to the p-shell nuclei, we emphasize that the Jacobi basis is inefficient and practically impossible to use. Instead, a basis of Slater determinants are employed, which are very easily constructed from single-particle states, once the appropriate antisymmetrization has been performed. The use of Slater determinants also allows for powerful second-quantization techniques to be easily employed in computer codes, since all the operations are performed on the single-particle level (i.e., is this state occupied or not?). However, it is exactly the afore mentioned convenience of operations on single-particle states that makes the computational cost of Slater determinant based codes prohibitive. As the size of the basis increases, the number of Slater determinants that are created in the $N_{\max}$ basis grows factorially. Since all computational resources are finite, we can store only so many basis states before we run out of resources. This is one of the reasons why the extrapolation to $N_{\max} = \infty$ has to be performed. In Fig. 2.7, I show the growth of Slater determinants as a function of $N_{\max}$ for several p-shell nuclei.

A further consideration is the storage of the Hamiltonian matrix elements. Most codes prefer to diagonalize the Hamiltonian matrix by means of the Lanczos algorithm. This algorithm is well-suited to nuclear-structure calculations, because we

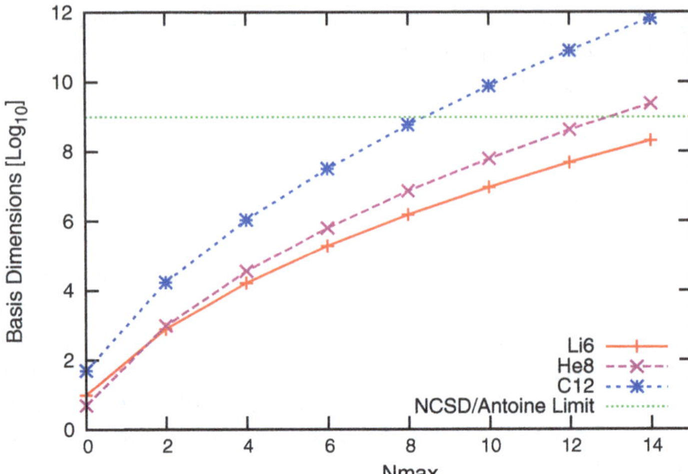

Fig. 2.7 The number of Slater determinants present in the m-scheme, as a function of $N_{\max}$ for various p-shell nuclei. The *horizontal line* shows the computational limit of two popular NCSM codes, which, in the case of using only two-body forces, corresponds to about one billion states

are usually interested in only the ground or a few low-lying states. However, we still need to store the matrix elements in memory. In the case of calculations that only involve two-body matrix elements, the Hamiltonian matrix (when expressed in Slater determinants) is sparse. For a p-shell nucleus at $N_{\max} = 14$ we need to store 5,481,920 non-zero two-body matrix elements (in the m-scheme). We will discuss how these matrix elements are stored in the next section. However, for three-body interactions, the number of non-zero matrix elements is much larger. The consequence is that we can only go up to about $N_{\max} = 8$ for p-shell nuclei when we include three-body forces. In the next section, I will address a recent development that somewhat overcomes this limit for three-body calculations.

2.5.1 Various NCSM Computer Codes in Use

The aim of this section is to give a brief overview of several computer codes used today, so that new researchers in the field can easily identify which resources are available and what some of the key differences are. All the codes listed below, except for the *Many Fermion dynamics* code have been used to produce the results in this thesis.

2.5.1.1 The No-Core Slater Determinant Code

The *No-Core Slater Determinant code (NCSD)* [49] is a code that has its roots in earlier versions of the *Many Fermion dynamics (MFD)* code [50, 51]. However, many changes have occurred since the early 1990s, making *NCSD* quite different to the current version of *MFD*. The version of *NCSD* that I describe corresponds to the July 2010 version. The code is multi-processor based, written in Fortran 90/95 and uses MPI for the parallelization routines. The code implements the m-scheme basis, as discussed earlier. Several distinct features are described below.

The code uses the Lanczos algorithm to determine the eigenenergies (as well as the wavefunctions) of the nuclear system. The Lanczos algorithm is particularly well-suited for many-body applications, especially if one is only interested in the lowest eigenvalues of the system (i.e., the gs). A brief summary of the Lanczos algorithm is that repeated applications of the Hamiltonian matrix produces a set of orthogonal Lanczos vectors, which transforms the original Hamiltonian matrix into a tri-diagonal form. After ≈50–150 Lanczos operations, the tridiagonal matrix is diagonalized by standard routines, leading to the low-lying states of the nucleus, in which we are interested. By far the most expensive part of the NCSM calculations are due to the repeated application of the Hamiltonian on the Lanczos vectors. The action of the two-body part of the Hamiltonian is shown below.

$$\sum_{ij,kl} a_i^\dagger a_j^\dagger a_l a_k \langle ij|H|kl\rangle \tag{2.24}$$

In Eq. (2.24) a summation over four indices needs to be performed. In computer codes, such a procedure can become very expensive, since the loop effectively scales as n^4, where n is the size of the basis. In *NCSD* this loop over four indices is reduced to a loop over only two indices, improving the scalability of the code. The two new indices do not correspond to the indices $ijkl$; instead we determine all combinations of annihilation operators, and determine all possible combinations of creation operators, in which the matrix element between the final ($\langle ij|$) and initial state ($|kl\rangle$) has the property that $\Delta M = 0$. In other words, we conserve the total angular momentum projection. Furthermore, we also check that the parity and isospin projections are conserved, further reducing the possible matrix elements that need to be looked up. This brings us to another point: determining which pairs of annihilation and creation operators lead to non-zero matrix elements is a good start, but we also need to look up the actual matrix element $\langle ij|H|kl\rangle$. In the case of *NCSD*, the final basis state $\langle ij|$ is looked up using a hash-table implementation. Hash-tables are particularly well-suited for such an application. The actual number of Slater determinants in a large NCSM calculation can be on the order of a few hundred million states. We do not want to search through a list containing millions of states, in order to determine which final states are required. An ideal hash-table would require exactly one lookup to determine which final state is required. In practical cases one deals with hash-table collisions, which require searching through a sub-list of about 10–40 final states to determine, which final state is needed.

2.5.1.2 Manyeff

Before any of the NCSM calculations are performed, one needs to generate the two- or three-body matrix elements of a specified interaction. In the case of the Machleidt and Entem EFT interaction, the momentum-space diagrammatic terms have to be expressed in the HO basis. The procedure is achieved by means of performing a Fourier transform from momentum to position space. Similar procedures are performed for other common interactions. The point we need to consider is that these interactions need to be expressed appropriately for the NCSM basis. In this regard, Petr Navrátil developed a code, commonly referred to as *manyeff* [39, 52], in order to produce the required two- and three-body Hamiltonian matrix elements for the NCSM.

The *manyeff* code eventually grew into a stand-alone NCSM code, capable of generating the matrix elements of a variety of interactions. Perhaps more interestingly, it is a code that is based on the use of relative (Jacobi) coordinates. Thus, *manyeff* became the tool that was used to perform extremely large $N_{\max}$ calculations of the Triton and ^{4}He. In the case of the Triton, it is now possible to calculate up to $N_{\max} = 40$ using *manyeff*, whereas for the Slater determinant based *NCSD* code, one can only go up to $N_{\max} = 16$. Naturally, when the generation of SRG interactions came along, the SRG procedures were built into *manyeff*, and were subsequently tested using the same code.

The code is written in Fortran 90/95 and is typically compiled with Intel compilers. *manyeff* has become a very powerful multiprocessor code, using both MPI and OpenMP multiprocessor routines in a hybrid setting. Many of the new capabilities, such as the SRG procedure as well as the multiprocessor capabilities were introduced by Eric Jurgenson [31].

2.5.1.3 ANTOINE

ANTOINE [53] is one of the older shell-model codes presently in use. Initially the code started as a traditional shell model code, meant for calculating heavier isotopes than what are typically calculated in the NCSM (i.e. $A > 16$). However, recent developments by Caurier and Nowacki, the creators of *ANTOINE*, have made NCSM calculations possible with the code. The two-body matrix elements are supplied by Petr Navrátil's *manyeff* code. *ANTOINE* is written in FORTRAN 77 and typically runs on a single processor. *ANTOINE* is based on the ideas of the Glasgow m-scheme method, but has some interesting improvements.

One such development is how Slater determinants are created in the code [54]. Typically, an entire Slater determinant consisting of proton and neutron single-particle states is created and stored in memory. In *ANTOINE*, separate proton and neutron

Slater determinants are created (this is also true for the 2012 version of *NCSD*). This procedure leads to a faster basis generation and requires less computer memory to store all the basis states. However, there is a slight complication: one needs to construct all proton and neutron Slater determinants that are possible in the single-particle basis, with all possible $m_{j,p/n}$ values (p/n corresponds to proton or neutron subspaces, respectively). Let us label the subspace Slater determinants for the protons as $|i_p, m_{j,p}\rangle$, where i_p enumerates the proton subspace state, and the neutron subspace Slater determinant as $|j_n, m_{j,n}\rangle$, where j_n enumerates the neutron subspace state. The product of the two subspace Slater determinants are combined to form a single Slater determinant $|i, JM\rangle$, in which $M = m_{j,p} + m_{j,n}$. The product of the proton and neutron Slater determinants can be written as

$$|i, M\rangle = |i_p, m_{j,p}\rangle |j_n, m_{j,n}\rangle$$

For example in the construction of a $M = 0$ Slater determinant, a $m_{j,p} = -1/2$ and $m_{j,n} = +1/2$ Slater determinant can form an $M = 0$ state; thus one must create all the Slater determinants with various $m_{j,p/n}$ in the proton and neutron subspaces in order to construct the complete list of $|i, M\rangle$ Slater determinants for a given M.

2.5.1.4 Many Fermion Dynamics

The *Many Fermion Dynamics* code (*MFDn*), is the current state of the art NCSM code and is written in Fortran 90/95 [55, 56]. The main developers of the code are James Vary and Pieter Maris, both at Iowa State University. *MFDn* is developed differently from all the other codes. Under the UNEDF collaboration, a team of computer scientists and applied mathematicians have added their input to the team, making it an extremely efficient piece of software. One particular development of note is the fact that the nuclear Hamiltonian is split up over many nodes on a large supercomputer, such as the Cray XT-4 at ORNL. This makes it possible to use terabytes of memory efficiently, allowing for the treatment of NCSM basis spaces on the order of 10 billion Slater determinants. Such an implementation would be useless, unless the code is load-balanced (which means that no single node is assigned more computations than any other node). The load-balancing has been implemented by the computer scientists working on the code. Two significant calculations have been performed with *MFDn*; there are predictions of the ^{14}F spectrum [57] as well as a study on the anomalous half-life of ^{14}C [58]. The theoretical predictions of the ^{14}F spectrum were confirmed 6 months later by an experiment performed at the Texas A&M Cyclotron [59]. There is, however, one comment that is in order regarding the current version of *MFDn*; there has been no implementation of three-body forces as of yet. In other words, *MFDn* is currently a two-body code only.

References

1. P. Navrátil, S. Quaglioni, I. Stetcu, B.R. Barrett, Recent developments in no-core shell-model calculations. J. Phys. G: Nucl. Part. Phys. **36**(8), 083101 (2009)
2. P. Navrátil, B.R. Barrett, Shell-model calculations for the three-nucleon system. Phys. Rev. C **57**, 562–568 (1998)
3. P. Navrátil, J.P. Vary, B.R. Barrett, Properties of ^{12}c in the Ab Initio nuclear shell model. Phys. Rev. Lett. **84**, 5728–5731 (2000)
4. P. Navrátil, V.G. Gueorguiev, J.P. Vary, W.E. Ormand, A. Nogga, Structure of $a = 1013$ nuclei with two- plus three-nucleon interactions from chiral effective field theory. Phys. Rev. Lett. **99**, 042501 (2007)
5. B. Alex Brown, W.A. Richter, New "USD" hamiltonians for the *sd* shell. Phys. Rev. C **74**, 034315 (2006)
6. V.G.J. Stoks, R.A.M. Klomp, C.P.F. Terheggen, J.J. de Swart, Construction of high-quality NN potential models. Phys. Rev. C **49**, 2950–2962 (1994)
7. Hideki Yukawa, On the interaction of elementary particles. i *. Prog. Theoret. Phys. Suppl. **1**, 1–10 (1955)
8. R. Machleidt, High-precision, charge-dependent bonn nucleon-nucleon potential. Phys. Rev. C **63**, 024001 (2001)
9. R.B. Wiringa, V.G.J. Stoks, R. Schiavilla, Accurate nucleon-nucleon potential with charge-independence breaking. Phys. Rev. C **51**, 38–51 (1995)
10. P. Doleschall, Influence of the short range nonlocal nucleon–nucleon interaction on the elastic $n - d$ scattering: below 30 MeV. Phys. Rev. C **69**, 054001 (2004)
11. P. Doleschall, I. Borbély, Z. Papp, W. Plessas, Nonlocality in the nucleon-nucleon interaction and three-nucleon bound states. Phys. Rev. C **67**, 064005 (2003)
12. J. Gasser, H. Leutwyler, Chiral perturbation theory to one loop. Ann. Phys. **158**(1), 142–210 (1984)
13. J. Gasser, H. Leutwyler, Chiral perturbation theory: expansions in the mass of the strange quark. Nucl. Phys. B **250**(1–4), 465–516 (1985)
14. S. Weinberg, Phenomenological lagrangians. Physica A **96**(12), 327–340 (1979)
15. S. Weinberg, Nuclear forces from chiral lagrangians. Phys. Lett. B **251**(2), 288–292 (1990)
16. S. Weinberg, Effective chiral lagrangians for nucleon-pion interactions and nuclear forces. Nucl. Phys. B **363**(1), 3–18 (1991)
17. U. van Kolck, Few-nucleon forces from chiral lagrangians. Phys. Rev. C **49**, 2932–2941 (1994)
18. C. Ordóñez, L. Ray, U. van Kolck, Nucleon-nucleon potential from an effective chiral lagrangian. Phys. Rev. Lett. **72**, 1982–1985 (1994)
19. C. Ordóñez, L. Ray, U. van Kolck, Two-nucleon potential from chiral lagrangians. Phys. Rev. C **53**, 2086–2105 (1996)
20. D. Gazit, S. Quaglioni, P. Navrátil, Three-nucleon low-energy constants from the consistency of interactions and currents in chiral effective field theory. Phys. Rev. Lett. **103**, 102502 (2009)
21. W. Gloeckle, E. Epelbaum, U.G. Meissner, A. Nogga, H. Kamada, et al., Nuclear forces and few nucleon studies based on chiral perturbation theory (2003)
22. D.R. Entem, R. Machleidt, Accurate charge-dependent nucleon-nucleon potential at fourth order of chiral perturbation theory. Phys. Rev. C **68**, 041001 (2003)
23. E. Epelbaum, A. Nogga, W. Glöckle, H. Kamada, Ulf-G. Meißner, H. Witała, Three-nucleon forces from chiral effective field theory. Phys. Rev. C **66**, 064001 (2002)
24. S.K. Bogner, T.T.S. Kuo, A. Schwenk, Model-independent low momentum nucleon interaction from phase shift equivalence. Phys. Rep. **386**(1), 1–27 (2003)
25. S.D. Głazek, K.G. Wilson, Renormalization of hamiltonians. Phys. Rev. D **48**, 5863–5872 (1993)
26. S.D. Glazek, K.G. Wilson, Perturbative renormalization group for hamiltonians. Phys. Rev. D **49**, 4214–4218 (1994)
27. F.J. Wegner. Flow equations for hamiltonians. Phys. Rep. **348**(1–2), 77–89 (2001) [Renormalization group theory in the new millennium. II]

28. F. Wegner, Flow-equations for hamiltonians. Ann. Phys. **506**(2), 77–91 (1994)
29. S.K. Bogner, R.J. Furnstahl, R.J. Perry, Similarity renormalization group for nucleon-nucleon interactions. Phys. Rev. C **75**, 061001 (2007)
30. S.K. Bogner, R.J. Furnstahl, R.J. Perry, A. Schwenk, Are low-energy nuclear observables sensitive to high-energy phase shifts? Phys. Lett. B **649**(5–6), 488–493 (2007)
31. E.D. Jurgenson, Applications of the similarity renormalization group tothe nuclear interaction. Ph.D. Thesis (Advisor: R.J. Furnstahl) (2009)
32. E.D. Jurgenson, P. Navrátil, R.J. Furnstahl, Evolution of nuclear many-body forces with the similarity renormalization group. Phys. Rev. Lett. **103**, 082501 (2009)
33. S. Ôkubo, Diagonalization of hamiltonian and tamm-dancoff equation. Progress Theoret. Phys. **12**(5), 603–622 (1954)
34. K. Suzuki, S.Y. Lee, Convergent theory for effective interaction in nuclei. Progress Theoret. Phys. 64(6), 2091–2106 (1980)
35. H. Kamada, A. Nogga, W. Glöckle, E. Hiyama, M. Kamimura, K. Varga, Y. Suzuki, M. Viviani, A. Kievsky, S. Rosati, J. Carlson, S.C. Pieper, R.B. Wiringa, P. Navrátil, B.R. Barrett, N. Barnea, W. Leidemann, G. Orlandini, Benchmark test calculation of a four-nucleon bound state. Phys. Rev. C **64**, 044001 (2001)
36. E.D. Jurgenson, P. Navrátil, R.J. Furnstahl, Evolving nuclear many-body forces with the similarity renormalization group. Phys. Rev. C **83**, 034301 (2011)
37. A.F. Lisetskiy, B.R. Barrett, M.K.G. Kruse, P. Navratil, I. Stetcu, J.P. Vary, Ab-initio shell model with a core. Phys. Rev. C **78**, 044302 (2008)
38. M. Moshinsky, Transformation brackets for harmonic oscillator functions. Nucl. Phys. **13**(1), 104–116 (1959)
39. P. Navrátil, G.P. Kamuntavičius, B.R. Barrett, Few-nucleon systems in a translationally invariant harmonic oscillator basis. Phys. Rev. C **61**, 044001 (2000)
40. P. Navrátil, B.R. Barrett, Four-nucleon shell-model calculations in a faddeev-like approach. Phys. Rev. C **59**, 1906–1918 (1999)
41. P. Navrátil, B.R. Barrett, W. Glöckle, Spurious states in the faddeev formalism for few-body systems. Phys. Rev. C **59**, 611–616 (1999)
42. L. Trlifaj, Simple formula for the general oscillator brackets. Phys. Rev. C **5**, 1534–1539 (1972)
43. B.J. Cole, R.R. Whitehead, A. Watt, I. Morrison, Computationa methods for shell-model calculations. Adv. Nucl. Phys. **9**, 123–176 (1977)
44. P.K. Rath, A. Faessler, H. Muther, A. Watt, A practical solution to the problem of spurious states in shell-model calculations. J. Phys. G Nucl. Part. Phys. **16**(2), 245 (1990)
45. D.H. Gloeckner, R.D. Lawson, Spurious center-of-mass motion. Phys. Lett. B **53**(4), 313–318 (1974)
46. C. Forssén, J.P. Vary, E. Caurier, P. Navrátil, Converging sequences in the ab initio no-core shell model. Phys. Rev. C **77**, 024301 (2008)
47. P. Navrátil, E. Caurier, Nuclear structure with accurate chiral perturbation theory nucleon-nucleon potential: application to ^{6}Li and ^{10}B. Phys. Rev. C **69**, 014311 (2004)
48. P. Maris, J.P. Vary, A.M. Shirokov, Ab initio no-core full configuration calculations of light nuclei. Phys. Rev. C **79**, 014308 (2009)
49. P. Navrátil. No core slater determinant code. Unpublished (1995)
50. J.P. Vary, D.C. Zheng, The many-fermion-dynamics shell-model code. Unpublished (1994)
51. J.P. Vary, The many-fermion-dynamics shell-model code. Iowa State University, Unpublished (1992)
52. P. Navrátil, Manyeff code. Unpublished (1998)
53. E. Caurier, F. Nowacki, Present status of shell model techniques. Acta Phys. Pol. B **30**(3), 705 (1999)
54. E. Caurier, G. Martínez-Pinedo, F. Nowacki, A. Poves, A.P. Zuker, The shell model as a unified view of nuclear structure. Rev. Mod. Phys. **77**, 427–488 (2005)
55. P. Maris, M. Sosonkina, J.P. Vary, E. Ng, C. Yang, Scaling of ab-initio nuclear physics calculations on multicore computer architectures. Procedia Comput. Sci. **1**(1), 97–106 (2010)

56. P. Sternberg, E.G. Ng, C. Yang, P. Maris, J.P. Vary, M. Sosonkina, H.V. Le, Accelerating configuration interaction calculations for nuclear structure, in *Proceedings of the 2008 ACM/IEEE conference on Supercomputing*, SC '08, Piscataway, NJ, USA (IEEE Press, 2008), pp. 15:1–15:12
57. P. Maris, A.M. Shirokov, J.P. Vary, Ab initio nuclear structure simulations: the speculative ^{14}F nucleus. Phys. Rev. C **81**, 021301 (2010)
58. P. Maris, J.P. Vary, P. Navrátil, W.E. Ormand, H. Nam, D.J. Dean, Origin of the anomalous long lifetime of ^{14}C. Phys. Rev. Lett. **106**, 202502 (2011)
59. V.Z. Goldberg, B.T. Roeder, G.V. Rogachev, G.G. Chubarian, E.D. Johnson, C. Fu, A.A. Alharbi, M.L. Avila, A. Banu, M. McCleskey, J.P. Mitchell, E. Simmons, G. Tabacaru, L. Trache, R.E. Tribble, First observation of 14f. Phys. Lett. B **692**(5), 307–311 (2010)

Chapter 3
Importance Truncated No Core Shell Model

3.1 Computational Issues with the NCSM

The NCSM has been very successful at describing light nuclei ($A \leq 6$), and in some cases, has also been able to describe nuclei in the middle of the p-shell (see [1] for an extensive list of results). However, NCSM calculations in the middle or in the upper part of the p-shell ($A \geq 10$) become very difficult to perform. Currently, interesting nuclei such as the Carbon or Oxygen isotopes are beyond the capabilities of the NCSM. To extend our calculations to the start of the sd-shell, is an even more challenging task. It is possible to do some exploratory calculations for the start of the sd-shell,[1] in which $N_{\max} \leq 4$, but fully converged results will be out of reach for many years. We remind the reader that by fully converged results, we mean calculations which are free of the two NCSM parameters ($N_{\max}$ and $\hbar\Omega$).

In the case of the NCSM, the main difficulty encountered in extending the calculations to heavier nuclei comes from the relatively quick rise in the number of many-body basis states present in the $N_{\max}$ spaces. Recall that for heavier nuclei, the Slater determinant basis is preferred over the Jacobi basis, since the antisymmetrization is much easier to implement. The Slater determinant basis however leads to a basis containing many more basis states than the Jacobi basis. Since all our calculations are performed on large supercomputers, the problem becomes one of memory capacity. We need to store the data structures that contain the basis states, as well as store the non-zero matrix elements of the Hamiltonian. As $N_{\max}$ increases, not only does the number of basis states increase rapidly, but so do the matrix elements that need to be stored; all this information consumes memory very quickly. Most supercomputers are configured with multiple CPU cores present on a single node (e.g., 8 cores per node), and have a memory allocation of 16 Gb. On average, each core can access typically 2 Gb of memory. Alternatively, if a single core is used per node, then 16 Gb of memory can be accessed by the CPU core, but, such resource allocations are regarded as wasteful. One might also argue that NCSM calculations

[1] These were done by Alexander Lisetskiy in 2007 but were never published.

M. K. G. Kruse, *Extensions to the No-Core Shell Model*,
Springer Theses, DOI: 10.1007/978-3-319-01393-0_3,

© Springer International Publishing Switzerland 2013

may require a large amount of storage memory (e.g., hard drives) or a huge amount of CPU time. The former is never a problem; the largest calculation performed in this thesis required about 100 Gb of storage memory, whereas hundreds of terabytes are available. On the other hand, CPU time is generally readily available; at worst the calculations will take a bit longer than anticipated to complete. Even when larger computers are built, our desire to push the forefront of these calculations will ultimately be limited by the underlying memory computer architectures.

The question now is, is there a way to do NCSM calculations for heavier nuclei? One might argue that the use of softened interactions such as those generated by the SRG, might make an extrapolation to $N_{\text{max}} = \infty$ possible, even if using only ground-state (gs) energies generated from the spaces $N_{\text{max}} \leq 6$–8. In practice, such extrapolations are unreliable and typically significantly overbind the nuclear systems. In order to produce reliable and meaningful $N_{\text{max}} = \infty$ extrapolations, the NCSM calculations require the gs energies from larger N_{max} spaces, typically in the range of $10 \leq N_{\text{max}} \leq 14$.[2] Thus, the use of softened interactions does aid us in accelerating the convergence of the NCSM, but does not provide us with a complete solution to the basis dimension problem. A complement to softened interactions is required.

Continuing the line of thought presented in the previous paragraph, one could propose the following calculational scheme. We expect that the gs wavefunction of a typical nucleus has components that are mainly found in the lower-lying HO shells. In other words, the basis states found in the $N_{\text{max}} \leq 4$ spaces are the dominant components of the gs wavefunction. As N_{max} increases, we expect that a small subset of the relatively many basis states present are contributing to the gs wavefunction. In fact, from a naive picture, one might even argue that including the one-particle one-hole (1p1h) configurations in the higher lying N_{max} spaces, on top of an $N_{\text{max}} = 4$ gs wavefunction, could lead to a reasonable description of the gs properties a nucleus. But how would one choose certain configurations (i.e., basis states) over other configurations? Furthermore, if one now starts to truncate the basis spaces, what kind of physics information is lost, or worse, are any uncontrollable errors introduced into a truncated NCSM calculation? These questions, the proposed methods, and finally the consequences of such a method are the discussion of this chapter.

By now the strategy we plan to implement should be clear: we are proposing to perform NCSM calculations in which we truncate the larger N_{max} basis space, selecting relevant basis states according to a method motivated by physics. One such method that has been proposed, is the Importance-truncated No-Core Shell Model (IT-NCSM) [2, 3]. In the IT-NCSM, a small set of basis states is chosen from the full N_{max} space, using a procedure based on first-order multi-configurational perturbation theory. As we have mentioned previously, the larger N_{max} spaces can consist of close to a billion (or more) Slater determinants.[3] However, in the IT-NCSM, typically only 10–15 million states are kept in the larger N_{max} spaces. At these basis dimensions,

---

[2] This point was expressed both by Pieter Maris and Robert Roth at the workshop on *Perspectives on the NCSM 2012.*

[3] Only the *MFDn* code can efficiently handle basis dimensions over one billion states.

the calculations are fairly routine to perform and progress (computationally) fairly quickly. The details of the IT-NCSM will be described in Sect. 3.2, in which we lay out the formalism of the method and the various improvements that can be made to the method.

The use of IT-NCSM in nuclear-structure calculations is a fairly new technique. It was first used by Robert Roth and Petr Navrátil in 2007 [2], and later refined by Roth in 2009 [3]. Currently, only the Roth group in Germany (TU Darmstadt) make use of IT-NCSM in their calculations. Thus, we find it appropriate to include a small subsection on the history of IT-NCSM calculations (see Sect. 3.3), as well as highlighting the differences to our approach.

The IT-NCSM calculations that have been published up to now, have clearly demonstrated the feasibility of the method [4, 5]. In all cases tested so far, IT-NCSM calculations and full NCSM calculations (when they are possible to perform) have agreed reasonably well. However, skeptics in the nuclear-structure community have challenged the use of IT-NCSM as a reliable method to determine gs energies of nuclei, especially when IT-NCSM is used for heavier systems such as ^{16}O or ^{48}Ca [6, 7]. In part, these criticisms stem from two concerns: (1) exactly how large is the 'error' in IT-NCSM calculations when compared to full NCSM calculations and (2) do IT-NCSM calculations violate sacred principles of many-body theory i.e., size-extensivity?[4]

Unfortunately, a serious discussion on the errors introduced by the IT-NCSM calculations, such as those introduced in the (required) extrapolations that recover the gs energy of the full $N_{\max}$ space, have not been presented. Similarly, a discussion on the errors introduced solely by working in a truncated $N_{\max}$ space, leading to possible violations of the Goldstone linked-diagram expansion [8], which is intimately linked with the concept of size-extensivity, have so far only been eluded to. We plan to address the extrapolation errors in great detail in Sect. 3.4, in which we present a calculation of ^{6}Li IT-NCSM gs (and excited state) energies, in which we perform calculations up to the $N_{\max} = 14$ basis space. These calculations are compared to full NCSM calculations, from which we attempt to deduce the reliability of IT-NCSM calculations. Once we have discussed our method of determining the errors in IT-NCSM extrapolations, we turn our attention in Sect. 3.5 to the observations of the IT-NCSM as NCSM parameters are changed (e.g., changes in $\hbar\Omega$). Furthermore, we include a brief discussion on the effects of other operators, such as the RMS matter radius, in Sect. 3.6. The issues regarding the Goldstone linked-diagram expansion, will be discussed in Sect. 3.7.

3.2 The Formalism of IT-NCSM

The main ideas of importance-truncation, as used in the NCSM, will now be presented. As we have already stated, the procedure is based on multi-configurational perturbation theory, as originally conceived in the area of quantum chemistry

[4] The review article of [9] has a good discussion on the issues of size-extensivity.

[10–12]. However, there are differences to what is done in nuclear-structure versus what is done in quantum chemistry. These differences will be discussed in Sect. 3.3.

Let us begin with a brief introduction. At the heart of the IT-NCSM procedure, lies a parameter κ, that is directly related to the number of many-body basis states kept in a certain N_{max} model space. We conveniently name the parameter κ the *importance measure* of a basis state. Naturally, one would desire that κ is based on a physical argument, related to which basis states should be kept in a truncated N_{max} space. Furthermore, one would also desire that variations in κ lead to predictable results in say the gs energy of the nucleus. It is the purpose of the next few sections to illustrate the particular choice of κ that we use, and why it is a reasonable choice to use.

3.2.1 The Selection of Many-Body Basis States via the Importance Measure

Before we delve into the technical matters related to the formalism of IT-NCSM, we should ask ourselves the question: when should one start to truncate the N_{max} basis spaces? In other words, should one attempt to do full NCSM calculations using as many complete N_{max} basis spaces as are reasonably possible, and then truncate only the last one or two remaining N_{max} spaces? Or could one start the N_{max} basis truncation at a lower N_{max} already?

To be more precise, consider the case of ^{6}Li. We can easily do the full NCSM calculations up to $N_{\text{max}} = 10$, in which there are approximately 10 million basis states. The next two basis spaces, $N_{\text{max}} = 12, 14$, have on the order of 50 and 211 million states, respectively. These basis states are still computationally feasible, but they consume a fair amount of computer resources. However, in order to perform a meaningful extrapolation to an infinite basis result (i.e., $N_{\text{max}} = \infty$), we require the gs energies of the $N_{\text{max}} = 12, 14$ basis spaces. Returning to our argument, we might consider truncating only the $N_{\text{max}} = 12, 14$ basis spaces. However, in practice, we prefer to start the basis truncation at a smaller N_{max} value; typically $N_{\text{max}} = 6$, although this can be adjusted at will. The reason for truncating the basis at lower N_{max} spaces is two fold: (1) The dominant components of a gs wavefunction are situated in lower lying HO shells. Typically those basis states are found in the $N_{\text{max}} \leq 4$ spaces. (2) The IT-NCSM procedure is supposed to choose exactly those states that are dominant in the gs (or excited state) wavefunction, otherwise the method is not performing as it should. As a final comment, we would like the IT-NCSM to have a mechanism in place that would 'correct' a poor choice for the initial N_{max} basis space that is truncated (e.g., if it turns out that most basis states in the truncated $N_{\text{max}} = 6$ really are 'important', then the procedure should include many of the $N_{\text{max}} = 6$ basis states, even though we initially attempted to truncate the basis space).

The details of IT-NCSM will now be presented. Suppose that one wants to perform a series of large $N_{\max}$ calculations ($N_{\max} > 8$) for a given nucleus, but due to limitations set by current computer architectures (or resources), the full-space calculation, in which all the basis states in those $N_{\max}$ spaces are kept, is not possible. However, let us assume that an $N_{\max} = 4$ calculation is easily performed, in which one is able to calculate the ground state wavefunction, which we will denote as $|\Psi_{\text{ref}}\rangle$. As a first-order approximation to the $N_{\max} = 6$ wavefunction, we could estimate the amplitudes of the $N_{\max} = 6$ basis states if we could perform the diagonalization in $N_{\max} = 6$. Such an estimation is offered by first-order perturbation theory.

$$|\psi^{(1)}_{N_{\max}=6,\text{IT}}\rangle = \sum_{\nu \in N_{\max}=12} \frac{\langle\phi_\nu|W|\Psi_{\text{ref},N_{\max}=4}\rangle}{\epsilon_\nu - \epsilon_{\text{ref,sp}}}|\phi_\nu\rangle \tag{3.1}$$

In Eq. (3.1) we have explicitly denoted $|\psi^{(1)}_{N_{\max}=6,IT}\rangle$, as the approximate wavefunction of the full space $N_{\max} = 6$ wavefunction. The $|\phi_\nu\rangle$ are the $N_{\max} = 6$ many-body basis states. $|\Psi_{\text{ref},N_{\max}=4}\rangle$ is our previously calculated reference state, which in our example we assume is the ground-state wavefunction of the $N_{\max} = 4$ space. The two terms in the denominator, refer to the single-particle energy level of the corresponding label. In our implementation, we always take $\epsilon_{\text{ref,sp}}$ to be the lowest unperturbed energy configuration of the nucleus. In ^{6}Li, this corresponds to taking $\epsilon_{\text{ref,sp}} = 2 * \hbar\Omega$, since two valence nucleons occupy the $N = 1$ shell. We neglect the zero-point motion of the HO, since we only require the difference in energy of the single-particle states. Furthermore, $\epsilon_\nu = (6+2)\hbar\Omega$ for the basis states in $N_{\max} = 4$. This is a particular choice that we make and is known as the Møller-Plesset type of partitioning. There are other choices that one can make for the energy-denominator however, these do not necessarily have superior convergence properties over the simple Møller-Plesset partitioning [11].

Note that Eq. (3.1) requires the matrix elements of the perturbation operator, W. A convenient definition of the W operator is to split the initial Hamiltonian H into two pieces, namely $H = H_0 + W$. We define H_0 to be that part of the Hamiltonian operator, which only connects many-body basis states that lie in the space $N_{\max} =$ 0–4. In other words, H_0 does not connect basis states from our reference space to the $N_{\max} = 6$ space and satisfies the eigenvalue equation $H_0|\Psi_{\text{ref}}\rangle = \epsilon_{\text{ref}}|\Psi_{\text{ref}}\rangle$. The full Hamiltonian, H, does, however, connect basis states from $N_{\max} = 6$ to the reference space. Thus, we can rewrite Eq. (3.1), by replacing the W operator with the full Hamiltonian, H, as follows.

$$|\psi^{(1)}_{N_{\max}=6,\text{IT}}\rangle = \sum_{\nu \in N_{\max}=12} \frac{\langle\phi_\nu|H|\Psi_{\text{ref},N_{\max}=4}\rangle}{\epsilon_\nu - \epsilon_{\text{ref,sp}}}|\phi_\nu\rangle \tag{3.2}$$

Such a form is extremely convenient, since we do not need to calculate any other matrix elements, than those we already have to calculate for the Hamiltonian operator. Equation (3.2) indicates that the largest correction to the wavefunction is essentially determined by the amplitude of the corresponding $N_{\max} = 6$ basis state.

The amplitude of the basis state is in turn determined by the Hamiltonian matrix element between the $N_{\mathrm{max}} = 6$ basis state and the reference state. This leads us to define the importance measure of a basis state, κ_ν, to be

$$\kappa_\nu = \frac{|\langle\phi_\nu|H|\Psi_{\mathrm{ref},N_{\mathrm{max}}=4}\rangle|}{\epsilon_\nu - \epsilon_{\mathrm{ref,sp}}}. \tag{3.3}$$

We can now use the importance measure as a way to set a threshold limit, as to which basis states are included in the truncated $N_{\mathrm{max}} = 6$ space. A typical value for κ is on the order of a few 10^{-5}. If we now set the threshold value of the importance measure to some value, say $3 * 10^{-5}$, we only keep those basis states (ϕ_ν) in $N_{\mathrm{max}} = 6$ for which $\kappa_\nu \geq 3 * 10^{-5}$. Since some states have an importance measure lower than this threshold, we will discard those states, and, thus, start to truncate the $N_{\mathrm{max}} = 6$ space. In reality, the number of states discarded depends not only on the threshold value, but also on which N_{max} basis space is currently being evaluated. Typically, the largest N_{max} spaces are most heavily truncated, whereas in the first few N_{max} spaces, most basis states are kept. This observation agrees with our intuitive guess, that the components of the ground state are dominated by basis states found in the lower oscillator shells.

As an initial test of the IT-NCSM method, we can investigate the correlation between κ_ν and the actual size of the components of $|\Psi_{N_{\mathrm{max}}=6,\kappa}\rangle$. For convenience, we separate the contributions of $N_{\mathrm{max}} \leq 4$ and $N_{\mathrm{max}} = 6$ basis states as follows,

$$|\Psi_{N_{\mathrm{max}}=6,\kappa}\rangle = \sum_{i\in N_{\mathrm{max}}\leq 4} c_i|\phi_i\rangle + \sum_{\nu\in N_{\mathrm{max}}=6} c_\nu|\phi_\nu\rangle$$

in which basis states $|\phi_i\rangle$ correspond to basis states found in $N_{\mathrm{max}} \leq 4$ and basis states $|\phi_\nu\rangle$ correspond to basis states that exist in $N_{\mathrm{max}} = 6$. If there is a large correlation between κ_ν and $|c_\nu|$, then we can argue that the importance measure is selecting exactly those basis states that *are* dominant in the $N_{\mathrm{max}} = 6$ basis space. In Fig. 3.1, we show the correlation plot of κ_ν and $|c_\nu|$ for the gs wavefunction of ^{16}O in the $N_{\mathrm{max}} = 2$ space, when the reference state is taken as the $N_{\mathrm{max}} = 0$ (single) Slater determinant.

3.2.2 *Properties of Importance Truncation and a Posteriori Corrections*

The selection procedure of the many-body basis states that are kept in the importance truncation calculation is based on a first-order perturbation theory result (for the wavefunction in $N_{\mathrm{max}} = 6$). Using a particular value for κ, leads to a specific number of many-body basis states being kept, spanning the now incomplete $N_{\mathrm{max}} = 6$ space, in which we diagonalize the full Hamiltonian, H. The diagonalization results in a

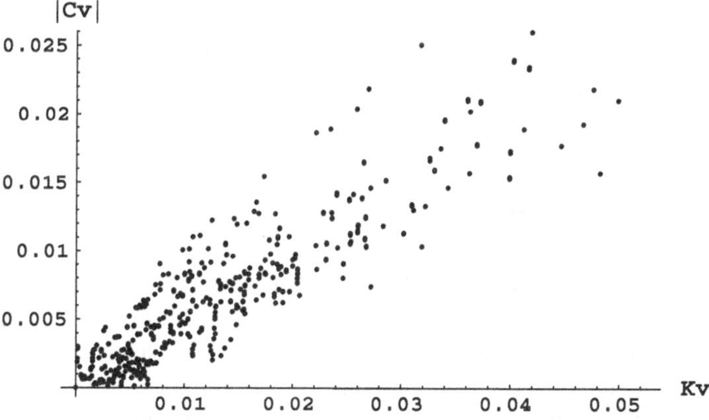

Fig. 3.1 The plot shows the correlation between κ_ν and $|c_\nu|$ for the gs wavefunction of ^{16}O in the $N_{\rm max} = 2$ space, when the reference state is taken as the $N_{\rm max} = 0$ (single) Slater determinant. The $N_{\rm max} = 2$ expansion coefficients c_ν of the wavefunction are compared to the value of κ_ν

ground state energy, $E^{(1)}_{0,\kappa}$, associated with the truncated wavefunction, $|\Psi_{N_{\rm max}=6,\kappa}\rangle$. Note that this wavefunction, which results from the diagonalization of the Hamiltonian in the truncated $N_{\rm max} = 6$ space, is not the same wavefunction $|\psi^{(1)}_{N_{\rm max}=6,\kappa}\rangle$, which is what we assume is a good approximation to the actual wavefunction, as shown in Eq. (3.2). Choosing a smaller value for κ, will result in more basis states being kept, a different truncated wavefunction, and thus will also result in a different ground state energy. By invoking the variational principle, we know that the calculated ground-state energy will decrease as we decrease the threshold value for κ.

It is also possible to estimate the energy contribution of the discarded states, by using second-order perturbation theory to determine the energy correction. The first-order energy correction vanishes, since we have defined the perturbation operator as $W = H - H_0$.

$$E^{(1)} = \langle\Psi_{\rm ref}|W|\Psi_{\rm ref}\rangle = \epsilon_{\rm ref} - \epsilon_{\rm ref} = 0 \tag{3.4}$$

In order to determine a non-zero quantity for the correction to the energy, we need to evaluate the second-order correction to the energy, by summing over all the discarded basis states, as shown below.

$$E^{(2)}_{0,\kappa} = - \sum_{\nu = {\rm discarded} N_{\rm max}=6} \frac{|\langle\phi_\nu|H|\Psi_{{\rm ref},N_{\rm max}=4}\rangle|^2}{\epsilon_\nu - \epsilon_{\rm ref,sp}} \tag{3.5}$$

Using this result, we can improve on the ground state energy calculated in the truncated space, $E^{(1)}_{0,\kappa}$ by adding $E^{(2)}_{0,\kappa}$ to it. The resulting energy,

$$E^{(1+2)}_{0,\kappa} = E^{(1)}_{0,\kappa} + E^{(2)}_{0,\kappa}, \tag{3.6}$$

has a smaller dependence on κ than $E^{(1)}_{0,\kappa}$ does, as will be illustrated later. However, $E^{(1+2)}_{0,\kappa}$ is no longer variational, as it is constructed from a quantity, $E^{(2)}_{0,\kappa}$, which is not a result obtained from the diagonalization of the Hamiltonian.

3.2.3 *The Extension to Excited States*

So far we have only discussed targeting basis states in the larger $N_{\max}$ space from a reference state, which we took as the ground state in a smaller $N_{\max}$ space. We can easily extend the basis selection procedure for excited states, by replacing the reference state with the desired excited state. In definite terms, this means that we define a κ threshold value, for each state in which we are interested, and evaluate all basis states in the larger $N_{\max}$ space for each reference state. The κ threshold value is taken as the same numerical quantity in each case. Returning to our previous example, in which we were selecting the basis states for $N_{\max} = 6$ from an $N_{\max} = 4$ reference state, we now define $\kappa^{(m)}_{\nu}$, as the corresponding κ for each of the m (ground and excited) states present. Note that the energy denominator remains the same for all the reference states used.

$$\kappa^{(m)}_{\nu} = \frac{|\langle\phi_{\nu}|H|\Psi^{(m)}_{\text{ref},N_{\max}=4}\rangle|}{\epsilon_{\nu} - \epsilon_{\text{ref,sp}}} \tag{3.7}$$

The number of basis states kept per fixed value of κ is larger for multiple reference states, when compared to the number kept when only one reference state is used. This is expected, since the structure of higher-lying states might be quite different to the ground state. The overlap between different reference states with the next-larger $N_{\max}$ basis states, $\langle\phi_{\nu}|H|\Psi^{(m)}_{\text{ref},N_{\max}=4}\rangle$, will be very different, depending on the structure of the relevant reference state.

3.2.4 *Implementation of Importance Truncation*

The Importance-Truncation procedure has been built into the *No-Core Shell Model Slater Determinant Code*, (*NCSD*) [13]. The relative simplicity of *NCSD* allowed for an easy modification of the code, so that the importance-truncation selection procedure could be done. We will now describe how it is done in our code, specifically for the reference state being the ground state (the extension to excited states is much the same).

The reader might have formed the impression that one needs to specify only one value of κ, in order to do 'the calculation'. Although it is certainly possible to calculate a good approximation to the next larger $N_{\max}$ space wavefunction by using just one value for κ, it is not sufficient to determine the actual energy that the complete N_{max} space would give. One needs to perform several calculations, for various κ threshold values, each resulting in a specific ground state energy, $E_{0,\kappa}$, so that an extrapolation to $\kappa = 0$ can be performed on the $E_{0,\kappa}$ values. The extrapolation to $E_{0,\kappa=0}$, which will be discussed in great detail in Sect. 3.4, yields what we assume is the true ground-state energy of the next-larger N_{max} space. Whether or not this is the case will also be addressed in Sect. 3.4, in which we compare extrapolated IT energies to the full-space NCSM energies, obtained from the *ANTOINE* code [14].

In order to calculate the series of $E_{0,\kappa}$ values that we need, an efficient algorithm was developed as follows. At the start of every calculation, we determine the smallest value of κ we would like to use. Most often, we choose the minimum value to be $\kappa_{min} = 1.0 * 10^{-5}$. Next, we construct *all* the basis states of the next-larger N_{max} space, and save those to a master file on disk. The master file is split up according to how many processors are used for the IT-NCSM calculation, typically ranging from 768 to 1536 processors. Each processor reads in the list of unique basis states ($|\phi_\nu\rangle$) assigned to it, and determines through one Lanczos iteration, which basis states satisfy the requirement that $\kappa_\nu \geq \kappa_{min}$. Those basis states that do satisfy this requirement are saved to a new file, along with the calculated value of κ_ν. This new file holds only a small fraction of the initial N_{max} many-body basis states; for the larger $N_{max} \geq 12$ spaces, the number of basis states kept are two to three-orders smaller than the full NCSM basis space.

Since we now have a list of all basis states that satisfy $\kappa_\nu \geq \kappa_{min}$, as well as their corresponding value for κ_ν, we are in a position to perform a series of calculations, in which we now vary κ. We define a series of κ values, for example, $\kappa = \{3.0, 2.0, 1.0\} * 10^{-5}$, and begin the calculation at the largest κ value. All states that now satisfy $\kappa_\nu \geq 3.0 * 10^{-5}$ are read in from the saved file, and are added to the many-body basis states already present. The resulting $E_{0,\kappa=3.0}$ energy is saved. The process repeats, in which we now add all the basis states that satisfy $\kappa_\nu \geq 2.0 * 10^{-5}$, that were not previously added. This procedure is repeated until we have calculated all the $E_{0,\kappa}$ for all the values of κ given. The resulting series of $E_{0,\kappa}$ values are then used to extrapolate to $E_{0,\kappa=0}$. Although we demonstrated the general procedure with only three κ threshold values, in general, we typically use 10–15 different κ threshold values.

The above procedure has been specific for one N_{max} space. In our calculations, we employ a bootstrapping idea, in which we apply importance truncation to several N_{max} spaces in a sequential order. This is very similar to the IT-NCSM(seq) technique of Roth [3]. We choose to begin with a complete $N_{max} = 4$ space, from which we construct the truncated basis in $N_{max} = 6$, using the appropriate reference state. Recall that this choice is arbitrary and can be changed by us if we desire. We then perform a series of calculations, as described above, in order to determine enough $E_{0,\kappa}$ values, for the $N_{max} = 6$ space, so that a reasonable extrapolation can be done to $E_{0,\kappa=0}$. Once we have calculated the energy for the smallest chosen κ value, we use that resulting wavefunction, $|\Psi_{ref,N_{max}=6,\kappa_{min}}\rangle$, as the reference state, for evaluating the $N_{max} = 8$ basis states. Besides checking *all* the basis states in $N_{max} = 8$, we also re-open the master list of *all* previously discarded basis states for $N_{\max} = 6$, and check if any of those states are *now* kept. This point will be discussed, along with other observations on the importance-truncation procedure, in Sect. 3.5. When we calculate the energy-contribution from the discarded basis states, we re-evaluate the contribution from all the states that were discarded up to the current point. For example, once we evaluate our $N_{\max} = 8$ basis states, we calculate the energy contribution of the discarded states from $N_{\max} = 8$ states as well as those that are still discarded in $N_{\max} = 6$.

$$E^{(2)}_{0,\kappa} = - \sum_{\nu=\text{discarded},\text{Nmax}=8} \frac{|\langle\phi_\nu|H|\Psi_{\text{ref},N_{\max}=6}\rangle|^2}{\epsilon_\nu - \epsilon_{\text{ref},\text{sp}}} - \sum_{\nu=\text{discarded},\text{Nmax}=6} \frac{|\langle\phi_\nu|H|\Psi_{\text{ref},N_{\max}=6}\rangle|^2}{\epsilon_\nu - \epsilon_{\text{ref},\text{sp}}} \tag{3.8}$$

This series of truncated $N_{\max}$ calculations continues until the desired $N_{\max}$ space is reached. To summarize, note that at the end of each series of calculations in a given $N_{\max}$ space, the wavefunction corresponding to the smallest κ value is used as the reference state for evaluating the basis states in the next larger $N_{\max}$ space, since that wavefunction is the best approximation (for the specified κ) to the complete $N_{\max}$ space. For each $N_{\max}$ space, we re-evaluate all the basis states that have been discarded in the lower $N_{\max}$ spaces, to check if any of them now satisfy the minimum κ threshold.

Before we turn our attention to extracting the ground state energies in IT-NCSM, we would like to demonstrate the effect on the size of the basis as κ varies. Recall that one of our principal goals was to reduce the size of the many-body basis to a computationally manageable size. In Fig. 3.2, we show the size of the basis in the IT-NCSM $N_{\max} = 14$ space (for ^{6}Li) as a function of κ. We show both the size of the basis when only a single reference state is used (the $J = 1^+$ gs) and the case when multiple reference states are used ($J = \{1^+, 3^+, 0^+\}$). We note that the basis in either case is on the order of a few million basis states; a significant improvement over the approximately 211 million basis states present in a complete $N_{\max} = 14$ space. Note that in the case of multiple reference states, the smallest κ threshold corresponds to $\kappa = 2.0 * 10^{-5}$, whereas in the single reference state the smallest value of kappa is $\kappa = 1.0 * 10^{-5}$. This is why fewer basis states were kept in the former case than in the latter, even though one might have expected the multiple reference state case to keep more basis states (this is the case when the smallest value of κ is actually the same). As a final comment on Fig. 3.2, we note that the sudden rise in the number of basis states kept at $\kappa \approx 1 * 10^{-5}$ is a sign that we are working in the appropriate range for κ. Once $\kappa < 1.0 * 10^{-5}$, we start to include many of the basis states that have a smaller amplitude in the total gs wavefunction. Figure 3.2 seems to suggest that at large values of κ, that the number of basis states kept is constant, or saturates. This reflects the case that we keep all basis states that were already included from $N_{\max} \leq 12$, regardless of what their importance measure might be in the current

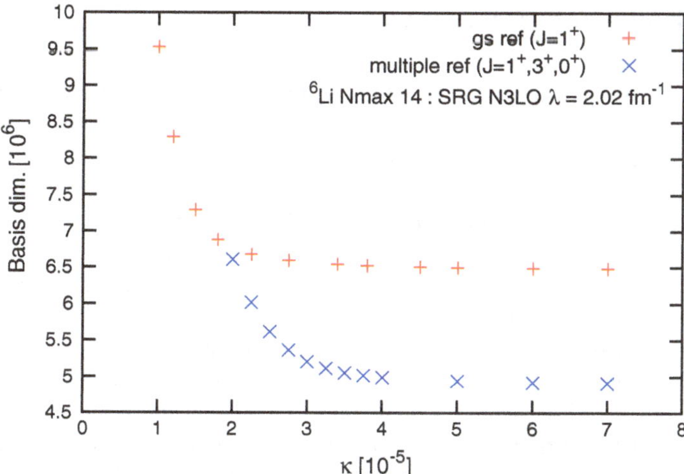

Fig. 3.2 The size of the (truncated) basis at $N_{\max} = 14$ in the case of ^{6}Li. Two *curves* are shown: the size of the basis for using a single reference state (+) or using multiple reference states (×). The full NCSM space at $N_{\max} = 14$ contains approximately 211 million states. The calculation was performed with the SRG chiral NN N3LO interaction, for which $\lambda = 2.02\,\text{fm}^{-1}$. The HO energy is $\hbar\Omega = 16\,\text{MeV}$

$N_{\max}$ space. In other words, once a basis state is kept, it is never discarded at a later stage.

At this stage, we have not presented any evidence that the basis states discarded by having $\kappa < 1.0 * 10^{-5}$ are not significant in determining the gs energy. We will return to this point in subsequent sections in which we will show that their contribution to the gs energy is minimal in comparison to including them in the truncated basis.

3.3 The History of IT-NCSM Calculations

The use of importance-truncation is reasonably new to nuclear-structure calculations, considering the first paper was first published in 2007. However, in the realm of quantum chemistry, the idea has been around since the late 1970s. We thus find it appropriate to discuss the historical developments of importance-truncation in nuclear-structure, specifically to address some of the changes that have occurred since the initial papers were published. The discussions of the development of the technique as a whole, particularly in the area of quantum chemistry, interesting as it is, is outside the scope of this thesis.

3.3.1 Particle-Hole Truncation

The first publication to appear on IT-NCSM calculations appeared in *Physical Review Letters* in August 2007 [2]. These calculations were presented by Robert Roth and Petr Navrátil and showed the gs energies for three nuclei, ^{4}He, ^{16}O and ^{40}Ca (all doubly-magic nuclei). These calculations were quite ambitious, even though they only included softened NN interactions (UCOM [15] and $v_{\text{low}-\text{k}}$), since the NCSM is typically not used to calculate the gs energies of nuclei as heavy as ^{16}O or ^{40}Ca. In this regard, the truncated N_{max} spaces made it possible to reach large enough N_{max} spaces to extract meaningful gs energies.

However, a different type of importance truncation scheme was employed than the one we outlined in Sect. 3.2. The particular importance truncation scheme was based on a particle-hole truncation method. Recall that the three nuclei we listed are considered to be doubly-magic, meaning that both the protons and neutrons completely fill the single-particle states up to a given major HO shell. From a many-body viewpoint, exactly one Slater determinant is required to describe the $N_{\text{max}} = 0$ configuration. This single Slater determinant is taken as the initial reference state, $|\Psi_{\text{ref}}\rangle$.

We will now proceed to describe what we now call the *iterative* IT-NCSM formalism. The presentation will be brief, so as to illustrate the main concept of this particular importance truncation scheme. As we mentioned before, the iterative scheme is based on a particle-hole truncation procedure, as we will now show. One begins with a general reference state,$|\Psi_{\text{ref}}\rangle$, in this case taken to the single Slater determinant that describes the $N_{\text{max}} = 0$ configuration of a doubly-magic nucleus. One would like to calculate the gs energy of the nucleus in a given N_{max} space; for simplicity let us assume the final basis space is $N_{\text{max}} = 4$, although in reality, it would be $N_{\text{max}} = 12$–16. Note that we specify only *one* N_{max} space in which to perform the calculation. The selection of the basis states in the chosen N_{max} space proceeds through the usual definition of the importance measure,

$$\kappa_\nu = \frac{|\langle\phi_\nu|H|\Psi_{\text{ref}}\rangle|}{\epsilon_\nu - \epsilon_{\text{ref,sp}}}.$$

Recall that basis states (ϕ_ν) that satisfy the criteria $\kappa_\nu \geq \kappa_{\text{min}}$ are kept in the truncated N_{max} space. In the case of a two-body Hamiltonian, the operation of the Hamiltonian H can promote two particles from the reference state (creating two holes in the process) to two unoccupied states in the $N_{\text{max}} = 4$ basis space (see the left side of Fig. 3.3). All possible two-particle two-hole (2p-2h) excitations are created, for which each possible configuration is evaluated according to the importance truncation threshold criteria. The basis states that do satisfy the threshold criteria are kept in the final truncated $N_{\text{max}} = 4$ basis space. The truncated 2p-2h $N_{\text{max}} = 4$ basis is then used to diagonalize the Hamiltonian, leading to a gs wavefunction $|\Psi_{\text{ref}}^{2p-2h}\rangle$.

However, there is a minor complication that must be discussed. We have in fact only evaluated *some* of the basis states in the full $N_{\text{max}} = 4$ basis space. Recall that

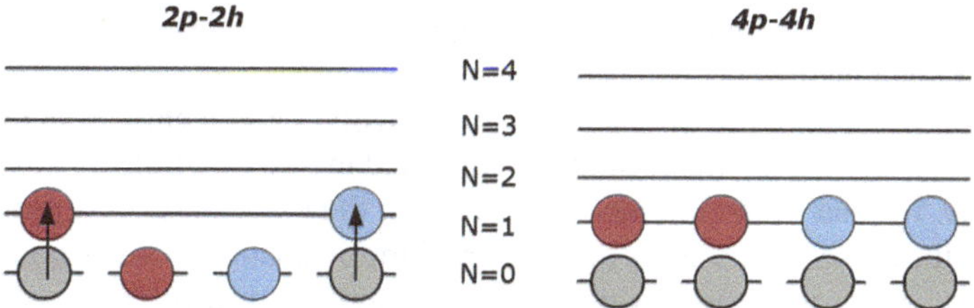

Fig. 3.3 On the *left*, we show one possible two-particle two-hole (2p-2h) excitation for ^{4}He, using the $N_{\max} = 0$ single Slater determinant as a reference state. For convenience, protons are indicated by the color *red*, whereas neutrons are colored *blue*. The *gray circles* correspond to the initial $N_{\max} = 0$ Slater determinant. On the *right*, we notice that the configuration shown is not evaluated in the 2p-2h importance truncation sampling, even though it is a basis state corresponding to $N_{\max} = 4$

an $N_{\max} = 4$ basis space implies that one has 4 quanta of HO energy available to share amongst the A nucleons. In our ^{4}He case, one possible $N_{\max} = 4$ configuration is shown on the right side of Fig. 3.3. This particular basis state was not evaluated in the 2p-2h truncation scheme. Thus, using $|\Psi_{\text{ref}}^{2p-2h}\rangle$ as a new reference state, one evaluates all the basis states again according to the threshold criteria. In effect, we have evaluated all 4p-4h configurations above the initial $N_{\max} = 0$ reference state (in the $N_{\max} = 4$ basis space). This gives us a new set of basis states, leading once again to a truncated $N_{\max} = 4$ basis space, except that this space now includes *all* possible configurations that could have been included from the $N_{\max} = 4$ space. The wavefunction that is determined from diagonalizing the Hamiltonian in this new 4p-4h basis is the final wavefunction that we desire.

The preceding few paragraphs illustrate several difficulties with the iterative approach to importance truncation. The main difficulty is that after one evaluation of a general $N_{\max}$ basis, one has really only evaluated the 2p-2h basis states. The procedure needs to be applied again in order to generate the 4p-4h basis states. Typically, the 4p-4h states that are now added to the 2p-2h basis states have a smaller effect on the gs energy, indicating that the 4p-4h states are 'less important'. Even then, depending on the size of A and $N_{\max}$, one might still not have evaluated *all* the basis states present in the general $N_{\max}$ basis space.

Furthermore, the truncation of basis states on a particle-hole level leads to size-extensivity issues. These will be discussed in Sect. 3.7.

3.3.2 Sequential $N_{\max}$ Calculations

The iterative (or particle-hole) importance truncation procedure was quickly abandoned. Although there are no serious errors in the method, the procedure was replaced by a much better implementation of importance truncation. In 2009, Roth published

[3] what can be considered as the most complete description of IT-NCSM and its various implementations.

The 'new' method of importance truncation involves a sequential scheme of evaluating basis states. In fact, the formalism presented in Sect. 3.2 follows this approach very closely. Instead of attempting to evaluate *all* the basis states of a general N_{max} space from the outset, one prefers to 'bootstrap' the evaluation of basis states sequentially in N_{max}. Let us assume we once again begin with a reference state in the $N_{max} = 0$ space. Considering that the Hamiltonian operator is a two-body operator, when it acts on the reference state, it will naturally generate $N_{max} = 2$ basis states. Once again, the importance measure κ_ν is used to evaluate all the $N_{max} = 2$ basis states. Note that we now generate *all* the $N_{max} = 2$ basis states from the action of the Hamiltonian on the reference state automatically. Once the truncated $N_{max} = 2$ basis is formed, we once again diagonalize the Hamiltonian in the truncated basis. This gives us a new gs wavefunction, $|\Psi_{\text{ref}}^{(N_{max}=2)}\rangle$, which we now use as a reference state to evaluate all the basis states in $N_{max} = 4$, using the same method as before.

The clear advantage of the sequential method, is that we now automatically generate all basis states in the evaluated N_{max} space. Thus, we only need to do it once per basis space. Furthermore, we can generate a sequence of gs energies for each truncated basis space, from which we can easily extrapolate the gs energy to $N_{max} = \infty$. Furthermore, there is *no* particle-hole truncation procedure in place. The question remains open whether or not the sequential IT-NCSM still violates size-extensivity (see Sect. 3.7).

3.3.3 Kruse and Roth IT-NCSM Implementations

In this section, we would like to discuss the differences between the implementations of IT-NCSM as is done in this thesis, and those as done by Roth. Since these differences are not well documented anywhere else, we find it appropriate to include them here.

Both Kruse and Roth use the sequential IT-NCSM formalism. Furthermore, the selection of the basis states always proceeds through the evaluation of the importance measure, κ_ν. There are essentially two subtle differences between the implementation of Kruse and that of Roth.

Sequence of Steps

The first difference arises in the steps that are followed to calculated the gs energies as a function of N_{max}. In the case of Kruse, one begins with the largest importance measure (κ) in a given N_{max} space, and then proceeds to calculate the gs energy for that particular set of basis states. The next gs energy that is calculated is for the next largest importance measure in the *same* N_{max} (truncated) basis space. The series

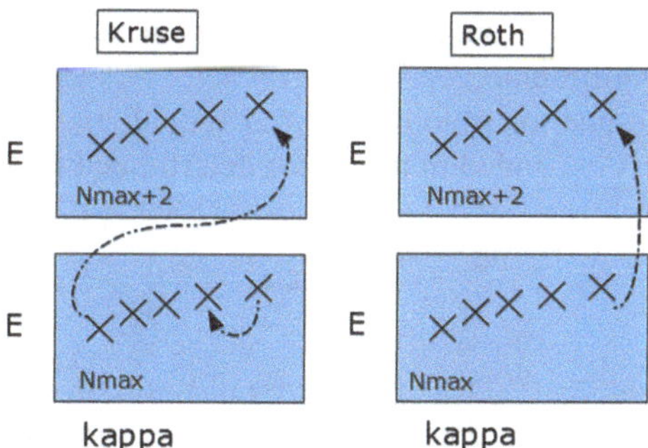

Fig. 3.4 The difference between how Kruse and Roth implement the series of steps in IT-NCSM calculations. In the case of Kruse (*left*), the approach is 'horizontal', in which all gs energies ($\times$) are calculated in the same $N_{\max}$ space before the next basis space ($N_{\max} + 2$) is evaluated. In the case of Roth (*right*) the approach is 'vertical'. A gs energy is calculated at a fixed value of κ for each $N_{\max}$ space. The process repeats anew with a smaller value of κ

of gs calculations continue until the smallest importance measure is reached. The resulting gs wavefunction that is determined from the $N_{\max}$ basis for $\kappa = \kappa_{\min}$ is used as the reference state for evaluating all the basis states in the next larger basis space ($N_{\max} + 2$).

In the case of Roth, κ is fixed initially. One then calculated the gs energies as a function of $N_{\max}$, while keeping κ fixed. Obviously, the wavefunction that is determined in each truncated $N_{\max}$ space is used to determine which basis states are kept in the next larger basis space. Once the calculation has completed up to the desired $N_{\max}$ space, the calculation begins anew, this time with a slighter smaller value for κ. The difference between Kruse and Roth is depicted graphically in Fig. 3.4.

This subtle difference has implications for the extrapolation of the gs energy (as a function of κ) in a given $N_{\max}$ space. Recall that we are interested in determining the gs energy at $\kappa = 0$. In Fig. 3.4, we have depicted gs energies as crosses ($\times$); these series of gs energies are extrapolated in various ways, the details of which will be presented in Sect. 3.4. We should point out though that the subtle difference between the two methods does affect the extrapolations. In the case of Kruse, the gs energies tend to vary more slowly as κ decreases in a given $N_{\max}$ space; in the case of Roth the gs energies tend to have a stronger dependence on κ (i.e., the slope of the curve is greater) for the same $N_{\max}$ space.

Truncation on the Reference State

There is one other difference between the IT-NCSM calculations of Kruse (this thesis) and those of Roth. In the sequential IT-NCSM scheme, a reference wavefunction for a given $N_{\max}$ space, $|\Psi_{\text{ref}}^{(N_{\max})}\rangle$, is used to evaluate the basis states in the next-larger

$N_{max} + 2$ basis space. The reference wavefunctions used between Kruse and Roth are, however, slightly different.

The initial gs wavefunction obtained from diagonalizing the Hamiltonian in the truncated N_{max} basis is expanded in the Slater determinant basis as follows,

$$|\Psi^{(N_{max})}\rangle = \sum_i c_i |\phi_i\rangle, \tag{3.9}$$

in which the $|\phi_i\rangle$ are the basis states. In the case of Roth, the reference wavefunction is obtained from $|\Psi^{(N_{max})}\rangle$, by imposing an additional truncation on the size of the expansion coefficients (c_i). The typical truncation threshold is set that only the basis components in $|\Psi^{(N_{max})}\rangle$ that satisfy $|c_i| \geq c_{min}$ are kept. The value of c_{min} is taken to be $c_{min} \approx 10 * \kappa$. Recall that in Roth's calculations, κ is fixed in each N_{max} space, so there is no 'minimum' value of κ used (i.e., κ is not necessarily the same as κ_{min}, as is the case in Kruse's calculations).

The actual reference wavefunction used by Roth is, thus,

$$|\Psi_{ref}^{(N_{max})}\rangle = \sum_{\nu, c_\nu \geq c_{min}} c_\nu |\phi_\nu\rangle. \tag{3.10}$$

The purpose of this procedure is to reduce the number of possible basis states that can be included from the $N_{max} + 2$ basis space. In effect, one limits the possible number of non-zero matrix elements of $\langle\phi_\nu|H|\Psi_{ref}^{(N_{max})}\rangle$ by reducing the number of basis states already present in the reference space. However, the argument is made that one only needs to keep the dominant terms in $|\Psi^{(N_{max})}\rangle$, in other words, those components that satisfy $c_\nu \geq c_{min}$, since it is those components that will have the most effect on the 'importance' of a newly added basis state from $N_{max} + 2$. The particular choice of having $c_{min} \approx 10 * \kappa$ is justified, by noting that there seems to be no noticeable difference in gs energies if the threshold is lowered.

The calculations done by Kruse *do not* use the c_{min} truncation on the wavefunction. In other words, $|\Psi^{(N_{max})}\rangle = |\Psi_{ref}^{(N_{max})}\rangle$. This choice is historical, but was later kept on philosophical grounds; do as little 'harm' to the reference wavefunction as possible.

3.4 Extrapolation Errors Introduced in the IT-NCSM

The errors that arise from working in a truncated basis have not been explored in depth. In order to form a clear picture in our minds about how various facets of the IT-NCSM function, we propose to do a series of extensive calculations, using ^{6}Li as our candidate nucleus. We choose this nucleus, as it one more complicated than the Triton or ^{4}He, but is not as challenging computationally as the mid p-shell nuclei. As a further consideration, we can perform full NCSM calculations for ^{6}Li all the way up to $N_{max} = 14$. This allows us to test the IT-NCSM in each of the N_{max} spaces that are truncated all the way up to rather large values of N_{max}. In most cases when the IT-NCSM is employed (e.g., heavier nuclei), one infers from the lower N_{max} spaces for which full NCSM calculations are still possible, that the method is performing very well. However, as N_{max} increases, the basis dimensions grow large, but the number of basis states kept in IT-NCSM only increases marginally. Thus, one may expect that the error in IT-NCSM calculations perhaps grows as N_{max} increases.

We begin with the bare NN N3LO interaction [16], which we transform to a phase-shift equivalent form, that has been evolved to momentum scales of $\lambda = 2.02\,\text{fm}^{-1}$. The choice of λ is fairly standard in terms of soft interactions. In later sections, we will vary λ to gain insights into the behavior of IT-NCSM as the character of the interaction changes.

As pointed out earlier, 10–15 different values of κ are used per N_{max} space, each resulting in a gs energy, $E_{0,\kappa}$. These are then used to extrapolate to $E_{0,\kappa=0}$. If $\kappa = 0$ in an IT-NCSM calculation, then all basis states are kept, thus, by extrapolating to $\kappa = 0$, we hope to recover the result of the complete N_{max} calculation. In this section, we will carefully analyze the extrapolation procedure and make a reasonable estimate of the error produced, simply by using various extrapolation techniques. Such an analysis is new and needs to be done. We will show that different conclusions can be drawn from the extrapolations, depending on how they were performed. It is not surprising to expect an error to be present in IT-NCSM calculations; however, we must point out that such results should be interpreted with care. One instance, where some care should be exercised, is in the extrapolation of a few N_{max} calculations to the infinite space ($N_{max} = \infty$). We will demonstrate that each calculated IT-NCSM N_{max} gs energy is associated with a small but finite error. These errors tend to grow as N_{max} increases. If an extrapolation to $N_{max} = \infty$ is now performed on these N_{max} points, one should expect an error to be associated with the predicted infinite result. The error on the infinite result is influenced by the finite N_{max} IT-NCSM calculations, and their respective errors. This naturally leads to the question, *how large is the associated error in the infinite result?*

In Fig. 3.5, we present a series of importance-truncated-calculated gs energies, for ^{6}Li in a truncated $N_{max} = 14$ space, using a range of κ values.

We have chosen 12 values of κ, given by the set $\kappa = \{7.0, 6.0.5.0, 4.5, 3.8, 3.4, 2.75, 2.25, 1.8, 1.5, 1.2, 1.0\} * 10^{-5}$. We will often refer to this set as the κ- grid points. This choice is arbitrary, although we did space the smallest κ values closer together, since we intuitively know that the smallest κ values have a larger effect on the extrapolation, than the larger κ values do. Our chosen range of κ, spanning from $\kappa = 7.0\text{–}1.0 * 10^{-5}$, is also to some extent arbitrary. Choosing a too narrow range could potentially affect the extrapolations in an undesired way, but choosing a too large a range, could bias the fitted functions too heavily towards $E_{0,\kappa}$ values that are associated with large values of κ. This brings us to our first point: The extrapolated values will depend on the chosen *range* of κ and will also be influenced by the *spacing* of the values of κ. We will address these issues in Sect. 3.4.1.

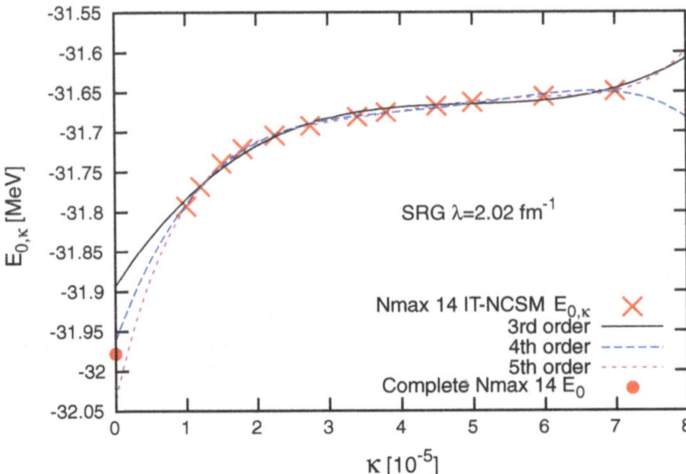

Fig. 3.5 IT-NCSM calculated gs energies of ^{6}Li in an $N_{\max} = 14$ space, using the SRG-N3LO potential with a momentum-cutoff $\lambda = 2.02\,\text{fm}^{-1}$. The oscillator energy is $\hbar\Omega = 16\,\text{MeV}$. The fit shows three different polynomial extrapolations to $E_{0,\kappa=0}$. Note that the extrapolated values are different and are spread across a range of about 150 keV. For comparison, we have included the gs energy of a complete $N_{\max} = 14$ space (*solid point*)

Another look at Fig. 3.5 suggests that the extrapolated values will also depend on the chosen function that is to be extrapolated. In Fig. 3.5, we present three possible choices; a third-, fourth- and fifth-order polynomial. The predicted gs energy in the $N_{\max} = 14$ space has a range of about 150 keV, between the third- and fifth-order polynomial. *A priori*, there is no class of functions that should be used in the extrapolations to the full-space result. Furthermore, we cannot make use of the Hellman-Feynman theorem, since the Hamiltonian does not explicitly depend on κ. In other words, there is no strict requirement that the extrapolated function should have a zero-derivative at $\kappa = 0$. This brings us to our second point: The extrapolated values also depend on which *type* of function was used to perform the extrapolation. An analysis of various functions, will be presented in Sect. 3.4.2.

The results presented in Fig. 3.5 are variational. They are calculated from a certain number of basis states, which are determined from the importance-truncation selection procedure for a given value of κ, in which the full Hamiltonian is diagonalized. An alternative way to fit the importance-truncated energies is to make use of the second-order corrections to the energy, $E^{(2)}_{0,\kappa}$, as shown in Eq. (3.5). Formally, we know that as $\kappa \to 0$, both $E^{(1)}_{0,\kappa}$ (as shown in Fig. 3.5) and $E^{(1+2)}_{0,\kappa}$ should meet at the same extrapolated point. It is, thus, also possible to do a constrained extrapolation of these two curves, one involving only the first-order energies $E^{(1)}_{0,\kappa}$, the other including the second-order corrections, $E^{(1+2)}_{0,\kappa}$, in such a way that both curves meet at the same point when $\kappa = 0$. Such an extrapolation is shown in Fig. 3.6, using the same NCSM parameters as in Fig. 3.5. Although we have only shown the fit for the fifth-order

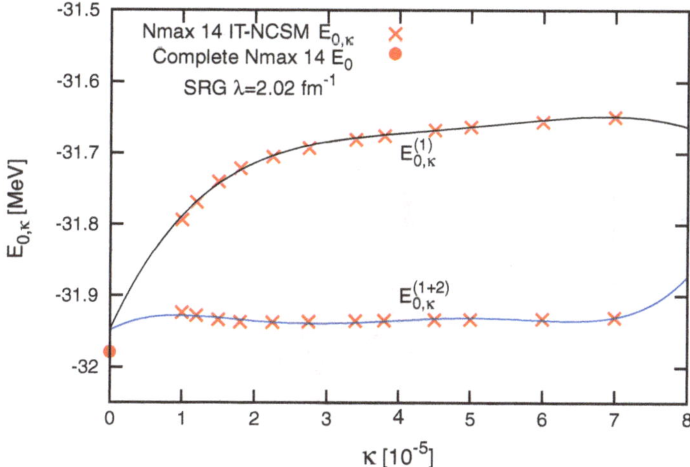

Fig. 3.6 A constrained fit on both the first-,$E^{(1)}_{0,\kappa}$ (*upper curve*), and second-order energies,$E^{(1+2)}_{0,\kappa}$ (*lower curve*), using a fifth-order polynomial. The NCSM parameters are the same as in Fig. 3.5 as well as the κ-grid

polynomial, it should be noted that using another polynomial will lead to a different extrapolated result. As will be shown later, when the constrained fit is used, the spread in extrapolated gs energies is lower than that suggested in Fig. 3.5. However, a spread in the extrapolated values does remain, and we would like to characterize how large that spread is.

3.4.1 Minimizing the Effect on the Chosen Set of κ Values

In Sect. 3.4, we had pointed out that the extrapolations to $E_{0,\kappa=0}$ depend on the chosen set of κ values. In particular, the extrapolation depends on the *range* of the set, the *number* of κ- grid points, as well as their *spacing*. In this section, we analyze the dependence on the extrapolated gs energy on these quantities. Such an analysis is quite interesting for the following reason: A different choice of κ-grid points leads to a different extrapolated value of the gs energy. Typically the range of κ is similar, spanning from a minimal value of a few 10^{-5} to at maximal value of about $20 * 10^{-5}$. In the larger $N_{\max}$ calculations, especially for the p-shell, it is computationally expensive to have $\kappa < 1.0 * 10^{-5}$, since the number of states grows exponentially when the value of κ is decreased.

The standard practice is to fit the 12 points shown in Fig. 3.5 by using some specified low-order polynomial. This, however, leads to one value of $E_{0,\kappa=0}$, without the ability to determine any error that is due solely to the extrapolation itself. As a first estimate of the error produced by varying ranges and spacings of the grid-points,

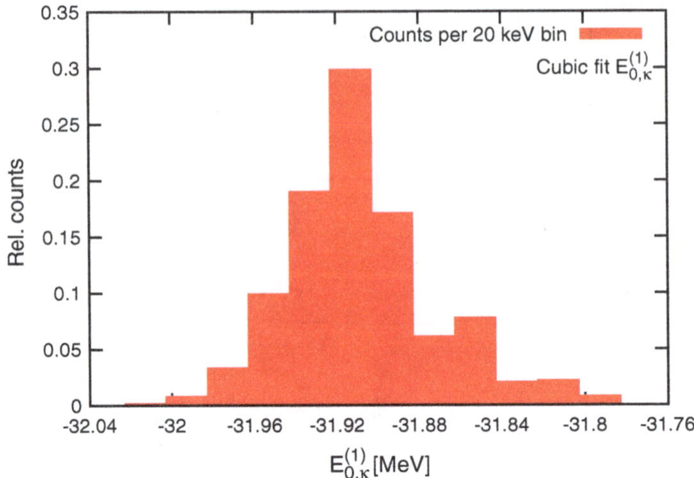

Fig. 3.7 The normalized distribution of extrapolated gs energies of ^{6}Li in the $N_{\max} = 14$ space, using $\hbar\Omega = 16\,\text{MeV}$. We use the chiral NN N3LO interaction, softened by SRG to $\lambda = 2.02\,\text{fm}^{-1}$. The extrapolations are done using a cubic polynomial, fitted only to the first-order energies, as was done in Fig. 3.5. The extrapolated values are binned by 20 keV. We determine the median to be $E_{0,\kappa=0} = -31.902\,\text{MeV}$ and the standard deviation to be $\sigma_{\binom{12}{7}} = 36\,\text{keV}$

as well as the number of grid points, we use the following procedure. We begin by choosing all possible combinations of 7 out of our 12 available points, and for each of these $\binom{12}{7} = 792$ sets, we fit a polynomial to and determine the extrapolated gs energy, $E_{0,\kappa=0}$. An example of the distribution of extrapolated energies, using a cubic polynomial fitted to the first-order energies, $E^{(1)}_{0,\kappa}$ is shown in Fig. 3.7. After calculating all the extrapolated gs energies that result from the 792 combinations of grid-points, we bin the results in 20 keV bins. From the distribution, we calculate the median as well as the standard deviation. We chose the median, instead of the average, as it is a statistical quantity that is not sensitive to outliers in the distribution. In the case of Fig. 3.7, we determine the median to be $E^{\binom{12}{7}}_{0,\kappa=0} = -31.902\,\text{MeV}$, which we also take as the value of the extrapolated gs energy. The standard deviation of the distribution shown in Fig. 3.7 is $\sigma_{\binom{12}{7}} = 36\,\text{keV}$.

The preceding paragraph lays the foundations of our error analysis. We repeat the above procedure for another 3 sets of data, created from choosing $\binom{12}{8}$, $\binom{12}{9}$ and finally $\binom{12}{10}$ combinations of κ grid points. For each data set, we determine the median of the distribution, as shown in Fig. 3.7, as well as the standard deviation. The median for each data set varies by at most a few keV, whereas the standard deviation decreases as the number of combinations of κ grid points decreases. We determine our extrapolated gs energy, as well as the associated error generated from the grid points by averaging over the calculated medians and standard deviations of the 4 data sets.

$$E_{0,\kappa=0} = \frac{E_{0,\kappa=0}^{\binom{12}{7}} + E_{0,\kappa=0}^{\binom{12}{8}} + E_{0,\kappa=0}^{\binom{12}{9}} + E_{0,\kappa=0}^{\binom{12}{10}}}{4} \tag{3.11}$$

$$\sigma = \frac{\sigma_{\binom{12}{7}} + \sigma_{\binom{12}{8}} + \sigma_{\binom{12}{9}} + \sigma_{\binom{12}{10}}}{4} \tag{3.12}$$

We should point out that our determination of the error, generated from various combinations of κ grid-points, with which we associate σ, is only a first attempt at determining the potential error of the extrapolations. The calculated standard deviation will, in general, differ if fewer (or more) data sets are used in Eq. (3.12). The important point we do want to make is that although the error might change depending on the number of data sets used, the order of magnitude of the error, whether it be a few or tens of keV's will not change. Such an estimate does have implications when extrapolations to $N_{\text{max}} = \infty$ are performed. We also note that the standard deviation is generally smaller, when one uses the constrained extrapolations, as is shown in Fig. 3.6.

3.4.2 Variations in the Chosen Function Used in the Extrapolation

In Sect. 3.4.1, we focused our attention on determining the error that is generated from various combinations of κ grid-points. The objective of that section was to average over many different possible choices of grid point configurations. In this section, we turn our attention to the choice of extrapolation function. As we had mentioned in the introduction to Sect. 3.4, there is no *a priori* justification to using one function over another. One simply goes by whether the chosen function, once fitted to the calculated $E_{0,\kappa}$, lies on top of the data or not. As can be seen from Fig. 3.5, various gs energies are predicted, depending on which function was chosen for the extrapolation. We will, thus, investigate various options that one might consider in fitting IT-NCSM calculated energies. We will use three different polynomials, a cubic, quartic, as well as a fifth-order polynomial.

For each selected function, we repeat the procedure outlined in Sect. 3.4.1. Besides for fitting the first-order results, $E_{0,\kappa}^{(1)}$, as in Fig. 3.5, we also repeat the fits, using the same function for both the first- and second-order results, $E_{0,\kappa}^{(1)}$ and $E_{0,\kappa}^{(1+2)}$ respectively, as shown in Fig. 3.6. Note that the constrained fits lead to a smaller standard deviation in the extrapolated gs energy.

3.4.3 Estimates of Extrapolation Errors

We will now present our calculated error estimates on the extrapolated gs energy in ^{6}Li, for the model spaces $N_{\text{max}} = 6$–14, and the oscillator value of $\hbar\Omega = 16$ MeV. The extrapolated gs energies are calculated, as well as the standard deviation,

Table 3.1 The extrapolated gs energies in various N_{max} spaces for ^{6}Li (NN N3LO SRG at $\lambda = 2.02\,\text{fm}^{-1}$) at $\hbar\Omega = 16\,\text{MeV}$, using only the first-order IT-NCSM calculated points, $E^{(1)}_{0,\kappa}$

N_{max}	NCSM	IT-NCSM	x^3 (MeV)	σ (keV)	x^4 (MeV)	σ (keV)	x^5 (MeV)	σ (keV)	Exact (MeV)
6	0.198	0.162	−28.601	≈ 0	−28.601	≈ 0	−28.601	≈ 0	−28.602
8	1.579	1.077	−30.216	2	−30.211	2	−30.211	2	−30.213
10	9.693	3.291	−31.207	2	−31.204	3	−31.197	4	−31.176
12	48.888	6.487	−31.714	10	−31.744	6	−31.741	21	−31.713
14	211.286	9.544	−31.899	25	−31.964	20	−32.046	29	−31.977

The table displays the various mean extrapolated values of the gs energy, for the polynomial used, as well as the calculated standard deviations of the fits, which are indicated to the right of the corresponding extrapolated energy. The exact result, in which all basis states are kept, is shown in the right-most column. The dimension (in millions) of basis states are shown in the complete N_{max} space as well as in the importance truncated space (columns 2 and 3). Note that the basis in IT-NCSM is drastically reduced in the larger N_{max} values

Table 3.2 The extrapolated gs energies in various N_{max} spaces for ^{6}Li (NN N3LO SRG $\lambda = 2.02\,\text{fm}^{-1}$) at $\hbar\Omega = 16\,\text{MeV}$, using the second-order IT-NCSM calculated points, $E^{(1+2)}_{0,\kappa}$

N_{max}	x^3 (MeV)	σ (keV)	x^4 (MeV)	σ (keV)	x^5 (MeV)	σ (keV)	Ex. (MeV)
6	−28.601	≈ 0	−28.602	1	−28.602	2	−28.602
8	−30.217	2	−30.211	2	−30.208	1	−30.213
10	−31.194	1	−31.196	1	−31.195	2	−31.176
12	−31.685	6	−31.702	5	−31.712	2	−31.713
14	−31.902	9	−31.925	12	−31.952	13	−31.977

The table displays the various mean extrapolated values of the gs energy, as well as the calculated standard deviations of the fits, which are indicated to the right of the corresponding extrapolated energy. The exact result, in which all basis states are kept, is shown in the right-most column

which we associate with the error generated from the extrapolation, as explained in Sect. 3.4.1. We also present the results from using various extrapolating functions. The extrapolated results are compared to the exact gs energies, as shown in Tables 3.1 and 3.2.

We observe the following trends presented by Tables 3.1, 3.2 and Fig. 3.8. The extrapolations to the exact gs energy for $N_{max} = 6$–8 are very good, being within one keV of the exact result and being independent of the function or method used. The agreement with the exact result is not surprising, as most of the many-body basis states are kept in those N_{max} spaces. Next, we observe that the error, σ, increases as N_{max} increases, but that it is smaller in the larger N_{max} spaces for the constrained second-order fits ($E^{(1+2)}_{0,\kappa}$), than the error for the corresponding first-order fits. Note that the error stated here is from variations in the combinations of κ grid points. This result is also expected, as now many basis states are discarded for the $N_{max} = 12$–14 spaces. However, note that for a given N_{max} space, the errors associated with each extrapolating function is roughly the same. This indicates that at least at some level,

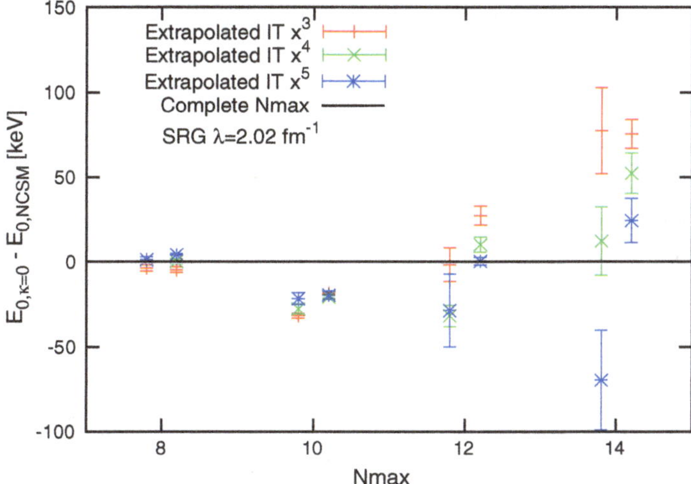

Fig. 3.8 The plot shows the extrapolated gs energy, relative to the exact gs energy (solid points), as well as the error, σ, that we determine from variations in the κ grid points. The points to the *left* are the extrapolations that are generated from fitting just the first-order set of data, $E^{(1)}_{0,\kappa}$. Those to the *right* are the extrapolations when fitting the second-order corrections, $E^{(1+2)}_{0,\kappa}$. The oscillator value is $\hbar\Omega = 16\,\text{MeV}$

choosing one function over another does not necessarily decrease the error from variations in the κ grid points.

We also note that the mean extrapolated gs energy, for a given N_{max} space, can be quite different between various functions, varying as much as 50 keV for the $N_{\text{max}} = 14$ space, when either the first- ($E^{(1)}_{0,\kappa}$) or constrained second-order ($E^{(1+2)}_{0,\kappa}$) results are fitted. The spread of extrapolated gs energies among the various chosen functions is usually quite a bit larger than the error associated with the variations in the κ grid points. Realistically, one does not have the exact calculations on hand, otherwise there would be no need for IT-NCSM, thus, characterizing the spread of the mean extrapolated gs energy is conceptually challenging. To illustrate this point, consider the results for $N_{\text{max}} = 12$ in Table 3.1. The cubic polynomial extrapolates to the exact result to within a keV, yet the quartic and fifth-order polynomial overestimate the result by about 30 keV, which is twice as large as the error from the variations in the κ grid points. If one does not know the exact result, one cannot make a reasonable guess, as to which functional extrapolation is the correct one to use. Furthermore, it should be clear from the tables that the error from variations in the κ grid points are lower than those associated with the use of different extrapolating functions. This leads us to state that the error from IT-NCSM extrapolations is *at least* as large as the stated value of σ, but *could be* as much as two- or three times larger. The latter is much harder to determine quantitatively.

Typically, the final procedure in any NCSM calculation is to extrapolate the gs energies as a function of N_{max} to $N_{\text{max}} = \infty$. The purpose of this procedure is to

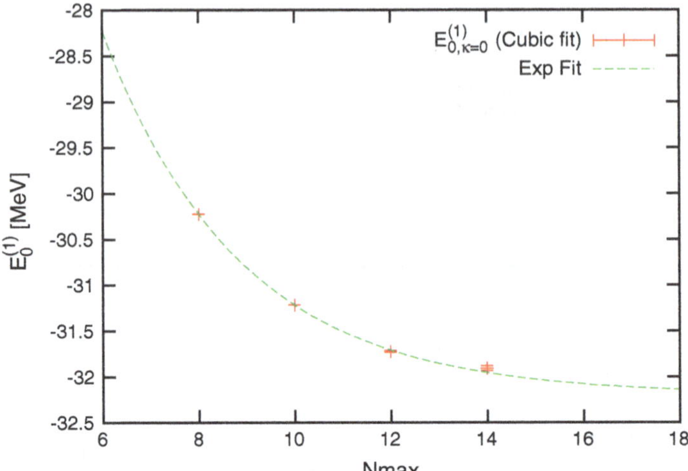

Fig. 3.9 The $N_{\max} = \infty$ extrapolation is shown, taking into account the σ error bars, in which we have used the first-order extrapolated results for $E^{(1)}_{0,\kappa=0}$, determined from fitting a cubic polynomial to the κ grid points. The function that we use is the commonly-used exponential decay, $a * \exp(-b * N_{\max}) + c$, in which c is the gs energy at $N_{\max} = \infty$. We determine the infinite result to be $E_{0,N_{\max}=\infty} = -32.188 \pm 0.031$ MeV, with a χ^2 per degree of freedom of 5.4. The NCSM model parameters are the same as before

remove the model parameters ($N_{\max}$, $\hbar\Omega$) from the calculations. The extrapolation to $N_{\max} = \infty$ removes the $N_{\max}$ dependence. The $\hbar\Omega$ dependence is removed, provided one chooses a HO energy near the variational minimum of the gs (as we have done). This procedure was first used in [17], in which the bare N3LO interaction was used to determine the gs energy of ^{6}Li.

In Fig. 3.9, we plot the extrapolation to $N_{\max} = \infty$, using the extrapolated values $E^{(1)}_{0,\kappa=0}$ that are determined from fitting a cubic polynomial to the first-order energies, $E^{(1)}_{0,\kappa}$. The extrapolation is done by using an exponential decay, $a * \exp(-b * N_{\max}) +$ c, in which we take into account the errors σ on the extrapolated points. We determine the infinite result to be $E_{0,N_{\max}=\infty} = -32.188 \pm 0.031$ MeV, with a χ^2 per degree of freedom of 5.4. We will compare this result to the result determined from the NCSM gs energies in the next paragraph. For now, let us point out that the $N_{\max} = \infty$ result has an error, σ_∞ associated with it. We determine $\sigma_\infty = 31$ keV. This value is not large, if we consider that the binding energy is on the order of 30 MeV.

We believe that the errors on the points shown in Fig. 3.9, arising solely from the distribution of the extrapolated values of $E_{0,\kappa=0}$, are due to variations in the κ grid points and are definitely present, but that the errors from other sources, which are much harder to determine, could make the *overall* error in IT-NCSM calculations larger than what we have calculated. Most importantly, we would like to emphasize the point that when considering IT-NCSM extrapolated results to $N_{\max} = \infty$, these extrapolations do come with some error, and hence, should at least be estimated.

Table 3.3 The $N_{max} = \infty$ extrapolations of the gs energy are shown

Method	x^3 (MeV)	σ_∞ (keV)	x^4 (MeV)	σ_∞ (keV)	x^5 (MeV)	σ_∞ (keV)
$E^{(1)}_{0,\kappa}$	−32.188	31	−32.334	26	−32.419	60
$E^{(1+2)}_{0,\kappa}$	−32.214	13	−32.207	15	−32.269	14

The extrapolations are performed as shown in Fig. 3.9, except now we have done it for both the first-order, $E^{(1)}_{0,\kappa}$ and second-order $E^{(1+2)}_{0,\kappa}$ points as well. The extrapolated $N_{max} = \infty$ gs energy, found from using the NCSM gs energies is −32.304 MeV

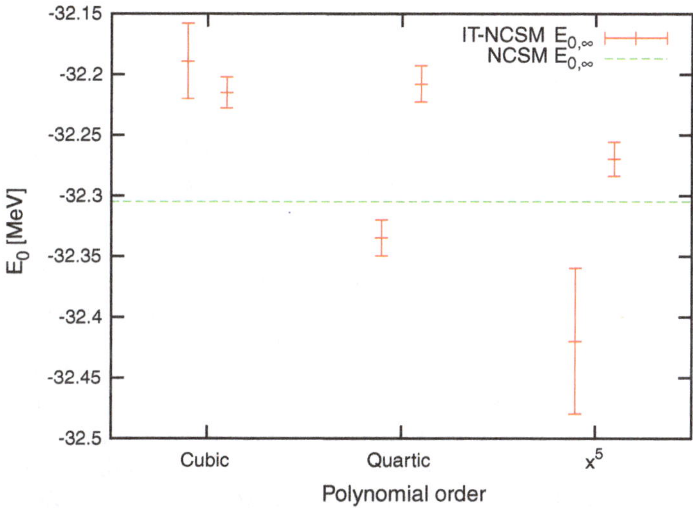

Fig. 3.10 The $N_{max} = \infty$ extrapolations of the gs energy are shown, as given in Table 3.3. Each polynomial order has two *lines* associated with it: to the *left* we show the extrapolation using the first-order results, and to the *right* the second-order results. The *horizontal line* indicates the result of extrapolating the full-space results to $N_{max} = \infty$. This value is −32.304 MeV

In Table 3.3 we show the resulting values of σ_∞ for other extrapolating functions, as well as using either the first- or second-order results. A graphical depiction of Table 3.3 is shown in Fig. 3.10, in which we also show the result of extrapolating full NCSM calculations to $N_{max} = 14$.

The results in Table 3.3 are encouraging, since they indicate that the error on $N_{max} = \infty$ extrapolations do not increase significantly. We also note that the difference from the exact $N_{max} = \infty$ extrapolation is on the order of tens of keV. We reiterate that, although this difference to the exact result is small in comparison to the gs energy, it is still present and could be relevant when conclusions are made from IT-NCSM $N_{max} = \infty$ extrapolated results.

3.5 Observations on the Importance Truncation Procedure

In the previous section, we addressed the fundamentals of IT-NCSM calculations. In particular, we addressed the choice of extrapolating function as well as the variation in κ-grid points. In this section we address further questions, namely, the dependence on the HO energy ($\hbar\Omega$), the SRG momentum-cutoff scale (λ), as well as the dependence of IT-NCSM gs energies on the number of reference states used (i.e., targeting excited states). These are discussed in Sects. 3.5.1, 3.5.2 and 3.5.4, respectively.

3.5.1 The Dependence on $\hbar\Omega$

All NCSM calculations have a dependence on the chosen HO energy, $\hbar\Omega$, even when bare interactions are used. However, the dependence on $\hbar\Omega$ can be minimized for a range of values. In practice, one typically chooses an $\hbar\Omega$ range resulting in the lowest gs energy of the largest $N_{\max}$ space employed. For our test case, we have included both optimal $\hbar\Omega$ choices ($\hbar\Omega = 16, 20\,\text{MeV}$), as well as two non-typical HO energies ($\hbar\Omega = 12, 24\,\text{MeV}$). As we will show later, there is a noticeable dependence on the quality of IT-NCSM calculations as $\hbar\Omega$ is increased.

Having determined the NCSM results, we can now determine if the IT-NCSM extrapolated results show any dependence on $\hbar\Omega$. In other words, regardless of the $\hbar\Omega$ value used, do we extrapolate to the corresponding NCSM result, or is there a systematic difference as a function of $\hbar\Omega$? In Fig. 3.11 we plot the difference of the extrapolated IT-NCSM gs energies, relative to the NCSM gs energy. Various extrapolating functions (cubic-, quartic- or fifth-order polynomials) are employed, using either the first- or second-order IT-NCSM extrapolated results, $E^{(1)}_{0,\kappa=0}$ or $E^{(1+2)}_{0,\kappa=0}$, respectively. From Fig. 3.11, we can see that there is a systematic drift away from the NCSM result as $\hbar\Omega$ increases. The discrepancy also increases as $N_{\max}$ increases, averaging about 200 keV from the exact result for the $N_{\max} = \infty$ extrapolations. In Sect. 3.5.3 we give some possible explanations for this type of behavior.

3.5.2 The Dependence on the SRG Momentum-Decoupling Scale (λ)

We now investigate the dependence on the SRG momentum-decoupling scale, λ. The NN chiral EFT N3LO potential, evolved to a momentum-decoupling scale of $\lambda = 1.5\,\text{fm}^{-1}$, has recently been used in the NCSM/RGM calculation of $d - \alpha$ scattering [18]. At this value of λ, the off-shell characteristics of the two-body potential have been changed in such a way, as to have the effect of producing a binding energy for ^{6}Li, similar to that obtained with the chiral EFT N2LO potential, in which the three-body terms are explicitly included. This behavior is due to the non-unitarity

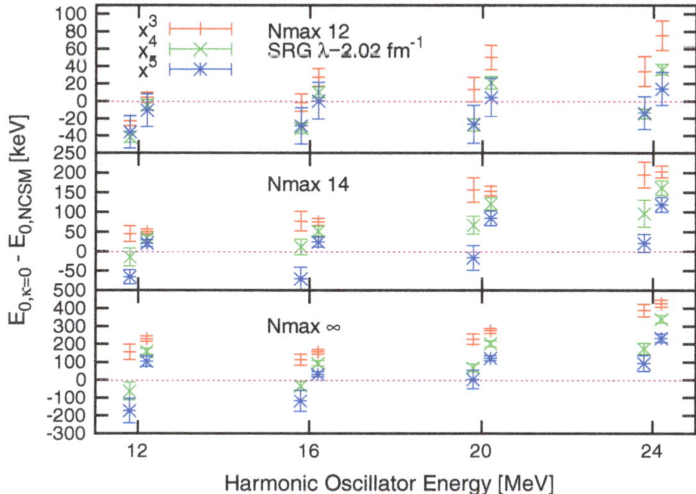

Fig. 3.11 The figure shows the relative difference of the IT-NCSM extrapolated energies, $E^{(1)}_{0,\kappa=0}$ (*left*) and $E^{(1+2)}_{0,\kappa=0}$ (*right*), to the NCSM gs energy (*horizontal curve*), as a function of the HO energy, $\hbar\Omega$. We also show the dependence on the various extrapolating functions, indicating the uncertainty in the extrapolation technique. Note that there is a systematic drift away from the NCSM result, as $\hbar\Omega$ increases. The discrepancy increases as $N_{\max}$ increases

of the SRG procedure, when only two-body terms are kept in the RG evolution. The effect for ^{6}Li is demonstrated in [19] (see Fig. 3.11). It is, thus, of interest to compare two different SRG evolved potentials, and to see if the importance truncation selection procedure behaves differently in the two cases. Lower λ's correspond to softer interactions, which translates into a faster convergence of the gs energy as $N_{\max}$ increases. It is interesting to check if IT-NCSM performs 'better' for softer interactions, or if there is no significant difference.

In Fig. 3.12 we plot the extrapolated IT-NCSM results, as a function of the HO energy, $\hbar\Omega$. This figure should be compared to Fig. 3.11 ($\lambda = 2.02\,\text{fm}^{-1}$). Note that the trends are similar, but the relative error to the NCSM results is a bit smaller for $\lambda = 1.5\,\text{fm}^{-1}$. The lower-$\lambda$ interactions are much softer, therefore, the convergence in $N_{\max}$ is much quicker. In this case, the IT-NCSM procedure selects fewer basis states for $\lambda = 1.5\,\text{fm}^{-1}$ than it does for $\lambda = 2.02\,\text{fm}^{-1}$, since for the softer potentials, fewer high-lying $N_{\max}$ basis states are required to reach convergence. This point is clearly illustrated in Fig. 3.13, in which we plot the number of basis states kept, as a function of $N_{\max}$ for both types of SRG evolved potentials.

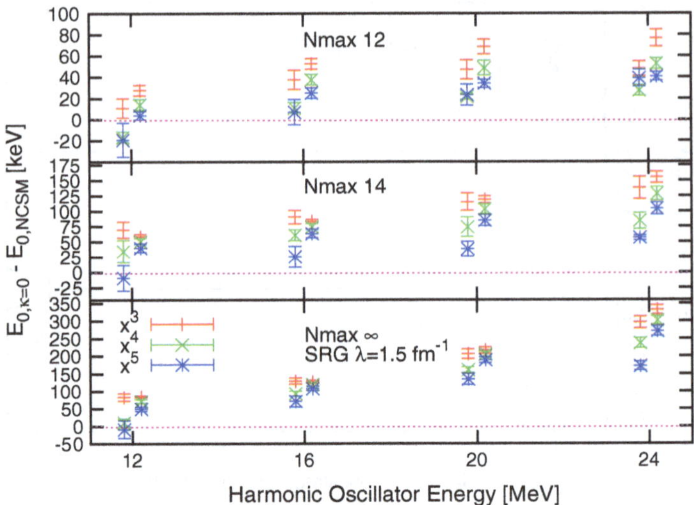

Fig. 3.12 The figure shows the relative difference of the IT-NCSM extrapolated energies, $E^{(1)}_{0,\kappa=0}$ (*left*) and $E^{(1+2)}_{0,\kappa=0}$ (*right*), to the NCSM gs energy (*horizontal curve*), as a function of the HO energy, $\hbar\Omega$. We also show the dependence on the various extrapolating functions, indicating the uncertainty in the extrapolation technique. Note that there is a systematic drift away from the NCSM result as $\hbar\Omega$ increases. A similar observation was made in Fig. 3.11

3.5.3 Further Comments on the $\hbar\Omega$ Dependence

Figure 3.13 shows another interesting trend; fewer basis states are kept as $\hbar\Omega$ increases. In particular, note that the basis is only about $\frac{3}{4}$ of the size for $\hbar\Omega = 24$ MeV that it is for $\hbar\Omega = 12$ MeV.

In Figs. 3.11 and 3.12, we noted that the IT-NCSM extrapolated results shift *away* from the NCSM results as $\hbar\Omega$ increases. We are now in a position to offer several explanations. The explanation lies in the definition of the importance measure κ. Recall that κ is inversely proportional to $\hbar\Omega$. Thus, for $\hbar\Omega = 24$ MeV, the matrix elements $|\langle\phi_\nu|H|\Psi_{\text{ref}}\rangle|$ would have to be twice as large as they are for $\hbar\Omega = 12$ MeV, in order for $|\phi_\nu\rangle$ to kept as a basis state. The matrix element itself also depends on $\hbar\Omega$, but this dependence must be weaker than the linear dependence in the denominator of κ (since fewer states are kept as N_{max} increases). Further evidence to the argument presented is shown in Fig. 3.14, in which we show the relative composition of various basis states, originating from various N_{max} spaces. The composition is determined by determining what percentage of basis states belonging to a certain N_{max} space (e.g., $N_{\text{max}} = 12$), make up the total basis in the final $N_{\text{max}} = 14$ basis space. Note that in Fig. 3.14, we clearly see that the number of basis states belonging to $N_{\text{max}} = 14$ in the total IT-NCSM basis are much smaller in comparison to those belonging to smaller N_{max} spaces. Recall that the last two or three N_{max} spaces are crucial in determining the extrapolation to $N_{\text{max}} = \infty$. Using the general argument we have presented, it

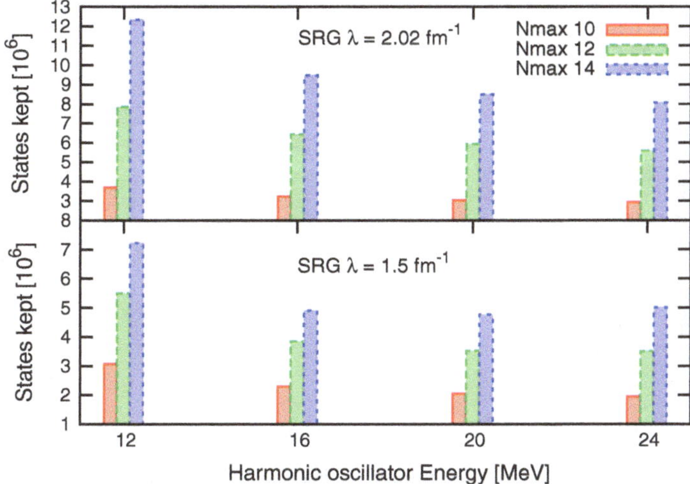

Fig. 3.13 The number of basis states kept, as a function of $N_{\max}$, for the $\lambda = 2.02\,\mathrm{fm}^{-1}$ (*top*) and $\lambda = 1.5\,\mathrm{fm}^{-1}$ (*bottom*) SRG evolved N3LO interaction. Note that for the softer interaction, $\lambda = 1.5\,\mathrm{fm}^{-1}$, IT-NCSM keeps fewer basis states from the larger $N_{\max}$ spaces. We also note that as $\hbar\Omega$ increases, fewer basis states are kept for the larger $N_{\max}$ spaces

is not too surprising that the IT-NCSM results would have a dependence on the HO energy and that, in general, the IT-NCSM results would be less reliable for larger values of $\hbar\Omega$ (if the same minimum value of κ is used), than for smaller values. We should also point out that, since the IT-NCSM basis is an incomplete $N_{\max}$ space (i.e., truncated), we no longer have a complete decomposition of center-of-mass from intrinsic states. Thus, a small amount of center of mass contamination is expected, which would increase as $\hbar\Omega$ increases. This issue has been addressed in [20].

3.5.4 Using Multiple Reference States

One final feature we would like to investigate is the behavior of the excited-state spectrum in IT-NCSM calculations. In order to reliably calculate the excited states of an IT-NCSM calculation, one needs to employ a reference state for each state that is to be calculated. In our case, we desire to calculate the gs ($J^\pi = 1^+$) and the first two excited states, corresponding to $J^\pi = 3^+, 0^+$. We, thus, use as initial reference states from $N_{\max} = 4$, each of those states, to generate the basis states that are kept in $N_{\max} = 6$. Since we are now using several reference states, the basis tends to be larger than if only one reference state is used. Thus, we have used a different set of kappa grid points, $\kappa = \{7.00, 6.00, 5.00, 4.00, 3.75, 3.50, 3.25, 3.00, 2.75, 2.50, 2.25, 2.00\} * 10^{-5}$. This range is a bit smaller than for a single reference state and only extends to $\kappa = 2.0 * 10^{-5}$, instead of $\kappa = 1.0 * 10^{-5}$, as before. In doing so, we attempt

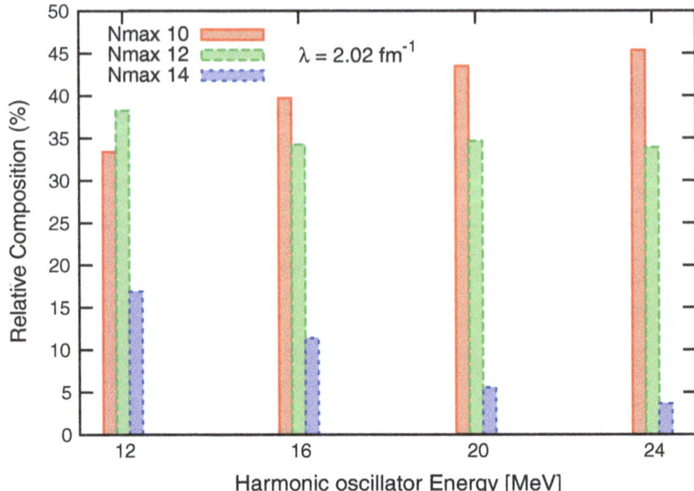

Fig. 3.14 The relative $N_{\max}$ composition of the basis as a function of $\hbar\Omega$, as determined in an IT-NCSM $N_{\max} = 14$ space. To be clear, the $N_{\max} = 14$ histogram refers to basis states that are found *only* in $N_{\max} = 14$ configurations, and so on. Note that the relative size of the $N_{\max} = 14$ basis size decreases as $\hbar\Omega$ increases. A similar trend is expected for $\lambda = 1.5\,\text{fm}^{-1}$

to include roughly the same number of states in both cases, so that we can form an opinion of which option really does select the better set of basis states. It might be that the inclusion of a different reference state (other than the gs) leads to basis states being kept in the truncated $N_{\max}$ space that were previously discarded.

In Fig. 3.15 we plot the relative difference to the NCSM result (horizontal line), for various extrapolating techniques, $E^{(1)}_{0,\kappa=0}$ (left) or $E^{(1+2)}_{0,\kappa=0}$ (right), as a function of increasing $N_{\max}$ ($\hbar\Omega = 16\,\text{MeV}$). This is to be compared with Fig. 3.8. The overall trend is the same between the two plots, indicating that the difference between IT-NCSM and NCSM calculations does not increase for the excited states, and that the difference is, in general, the same size as before (about 100 keV for $N_{\max} = 14$). In other words, the excitation spectrum can be calculated with the same degree of accuracy as for the gs.

An interesting comparison to make between using one or several reference states is to determine the behavior of the gs energy as a function of the number of basis states kept. This analysis is shown in Fig. 3.16, for $\hbar\Omega = 12\,\text{MeV}$. We chose that particular value of $\hbar\Omega$, since then the most number of basis states are kept. From the figure one can deduce two interesting points. The first point is that the additional reference states are selecting basis states that were previously not selected, as can be seen at the start of each $N_{\max}$ space. These additional basis states tend to make the functional dependence of the gs energy as a function of the size of the basis approximately constant. It can also be seen that when multiple reference states are used, fewer states are needed than before, in order to achieve the same gs energy as with a single reference state. The second point has to do with the lowering of the gs

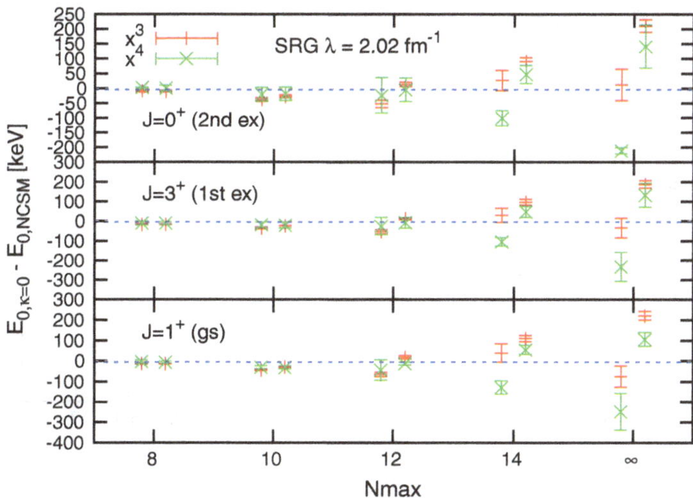

Fig. 3.15 The figure shows the relative difference to the NCSM result (*horizontal line*), for various extrapolating techniques, $E^{(1)}_{0,\kappa=0}$ (*left*) or $E^{(1+2)}_{0,\kappa=0}$ (*right*), as a function of increasing N_{max}. In this case, three reference states were used ($J^{\pi} = 1^{+}, 3^{+}, 0^{+}$). We calculate the gs and the first two excited states. Note that the overall trend is the same for all three states, indicating that IT-NCSM performs equally well for excited states as for the gs. We have also performed the extrapolation to $N_{\text{max}} = \infty$. ($\hbar\Omega = 16\,\text{MeV}$)

energy as a function of N_{max}. Note that higher N_{max} contributions significantly lower the gs energy, when compared to simply adding more states in a single N_{max} space. In other words, note that the drop in energy between $N_{\text{max}} = 12 \rightarrow 14$ is larger than the drop in energy in just the $N_{\text{max}} = 12$ space, resulting from adding all the basis states that are kept. Such a feature could hold promise for doing configuration-interaction calculations, in which one- and two particle-hole excitations are created on top of a Hartree-Fock state, generated in a small N_{max} space. Most of the binding energy is gained from adding the most notable configurations found in larger N_{max} spaces. However, we did notice in our IT-NCSM calculations that as N_{max} increases previously discarded basis states in the lower N_{max} spaces *do* become relevant at some stage and are added back into the basis by our basis evaluation procedure. This also explains why we typically use all states up to and including $N_{\text{max}} = 4$; those basis states will be added to the basis in any case by IT-NCSM. In fact, by the time we have completed our $N_{\text{max}} = 14$ calculation, almost all of the $N_{\text{max}} = 6$ states have been added to the basis, even though initially a fair number of those states were discarded at the start of the calculation. Such behavior makes sense, since we believe that states such as the gs, have components mostly found in the low-lying oscillator shells.

We expect much of the same results to hold for multiple reference states, as shown for the single reference state calculations. In particular, we have determined that the $\hbar\Omega$ dependence, as shown in Fig. 3.11, still persists when multiple reference states

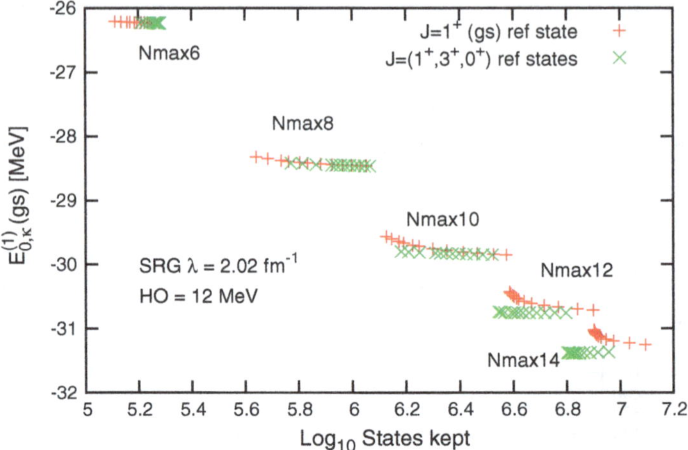

Fig. 3.16 The figure shows the gs energy at $\hbar\Omega = 12\,\text{MeV}$ as a function of the Logarithm (base 10) of the number of basis states kept. The *curve* with (+) signs corresponds to when only the gs is used as a reference state, wheres the *curve* marked ($\times$) shows the behavior when the lowest three states are used as reference states. Note that these energies correspond to the first-order results, $E^{(1)}_{0,\kappa}$, and that the two sets of kappa grid points are not identical. The figure shows that higher $N_{\max}$ contributions significantly lower the gs energy. Furthermore, using several reference states leads to an approximately constant dependence for the gs energy on the number of basis states kept

are used. This once again leads us to the conclusion that the dependence stems from κ being inversely proportional to $\hbar\Omega$.

3.6 General Operator Behavior in IT-NCSM

3.6.1 Introduction to General Operators

The study of operators, such as transition matrix elements or radii, in the IT-NCSM has so far not been explored. All published results up to now, have explicitly dealt with determining the gs or more generally, the energy spectrum. Needless to say, there are reasons for the omission of a detailed study of other operators besides for the Hamiltonian. To begin with, even in the NCSM, the energy spectrum is usually the most often calculated quantity, whereas operators like radii are usually casually mentioned or not presented at all. There are of course some notable exceptions [21–25].

General operators are a great deal more 'complicated' than the Hamiltonian, since they often converge more-slowly in $N_{\max}$. We explain why this is so. In order to understand the difficulties associated with general operators in the NCSM, we will use as an example the matter-radius operator, $\sqrt{\langle\Psi_0|r^2_m|\Psi_0\rangle}$, in which Ψ_0 represents the gs wavefunction. We define r^2_m as follows,

$$r_m^2 = \frac{1}{A}\sum_i^A (\vec{r_i} - \vec{R_{\rm CM}})^2 \tag{3.13}$$

in which we explicitly remove any contributions from the center of mass $\vec{R_{\rm CM}}$ coordinate. In the (IT-)NCSM, the gs wavefunction ($|\Psi_0\rangle$) has a Gaussian tail, i.e., the wavefunction falls off as $e^{-(\frac{r}{b})^2}$. However, bound-state wavefunctions actually have an exponential tail, meaning the true wavefunction falls off as e^{-cr}. If one is now interested in calculating an operator like the matter radius r_m^2, one very quickly notices that these types of operators converge slowly in $N_{\rm max}$. This is, of course, due to the incorrect asymptotic form of the wavefunction; increasing $N_{\rm max}$, increases the spatial extend of the gs wavefunction, but is, unfortunately, 'damped' by the Gaussian envelope. Operators that depend on r are typically referred to as long-range operators; the quadropole moment is another notorious example of a long-range operator.

There is another issue present, which one does not need to consider for the Hamiltonian. In the types of calculations that we are performing, in which we use a (softened) bare interaction for the nuclear interaction, we are guided by the variational principle. We know for a fact that increasing $N_{\rm max}$ will *always* lead to a lower (or constant) gs energy than before. Thus, the gs energy is a monotonically decreasing function of $N_{\rm max}$, regardless of the HO energy ($\hbar\Omega$) chosen. This justifies our extrapolation techniques for the gs energy. Recall that in the case of the gs energy, we use an exponential decay to extrapolate the gs energy to $N_{\rm max} = \infty$. General operators *do not* obey the variational principle [26]. It often is the case that for a range of HO energy values, the expectation value of an operator decreases as a function of $N_{\rm max}$, whereas for other HO energy ranges, the expectation value may slowly *increase*. In the case of effective interactions, both cases may be witnessed as $N_{\rm max}$ increases for the same HO energy. Each situation must be dealt with on a case-by-case basis.

Ultimately, the difficulties that were described in the preceding paragraphs, hamper our attempts to form a coherent and consistent approach to extrapolating general operators to $N_{\rm max} = \infty$.

Nevertheless, we now explore the behavior of r_m^2 in the IT-NCSM. It is the simplest long-range operator that we can study and is obviously closely connected to experimental results (although obtaining agreement with experimental data in this work is not our main goal). From a physics view-point, looking at other operators calculated from IT-NCSM is rather interesting. Recall that the truncated basis is ultimately selected by the matrix element $\langle\phi|H|\Psi_{\rm ref}\rangle$. In other words, the Hamiltonian is responsible for the selection of which particular basis states are included in the truncated basis. The question is now, does the Hamiltonian also select the basis appropriately for general operators? In contrast, one might argue that the current IT-NCSM does a good job for gs energies (as well as excited states), but for the case of a general operator $\hat{O}$, perhaps the selection of basis states specific for an optimization of $\hat{O}$ in the truncated basis should be based on an analogous expression, $\langle\phi|\hat{O}|\Psi_{\rm ref}\rangle$. We will only explore the former consideration and leave the latter as an open question for now.

3.6.2 The Matter-Radius in IT-NCSM ^{6}Li Calculations

As before, we will use ^{6}Li as our test nucleus. We will consider the behavior of the matter-radius as a function of $N_{\rm max}$ as well as for varying $\hbar\Omega$ values. Since we are now dealing with an operator other than H, we will need to develop a new extrapolation technique for both extrapolating to $\kappa = 0$ as well as for extrapolating to $N_{\rm max} = \infty$. The employed interaction will be the SRG NN N3LO interaction, in which $\lambda = 2.02\,{\rm fm}^{-1}$.

Extrapolating to a Complete $N_{\rm max}$ Space

Just like we did with the IT-NCSM gs energies, we will first fix the HO energy to $\hbar\Omega = 16\,$MeV. By fixing $\hbar\Omega$ we can show how we extrapolate to $\kappa = 0$ in various $N_{\rm max}$ spaces as well as present the final extrapolation to $N_{\rm max} = \infty$. This also gives us the opportunity to discuss how we determine the errors from the extrapolation procedures. As a reminder, we begin with a complete $N_{\rm max} = 4$ basis spaces. The basis spaces $N_{\rm max} = 6$–14 are all truncated according to our IT-NCSM procedure, as we described in Sect. 3.2.4.

In Fig. 3.17, we show the results for the matter radius, $\sqrt{r_m^2}$, as a function of κ, in the two largest (truncated) $N_{\rm max}$ spaces. We note that the points can be fitted by

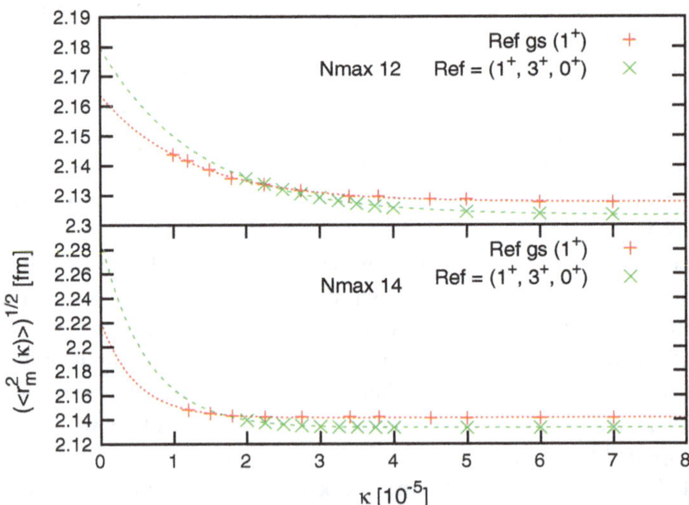

Fig. 3.17 The figure shows the extrapolations to $\kappa = 0$ for the matter radius, $\sqrt{r_m^2}$. We display the IT-NCSM results for $\sqrt{r_m^2}$, in which the reference state was either the gs (+) or multiple reference states (×). The extrapolation is given by the function $f(\kappa) = ae^{-\kappa b} + c$. Note that the value we are interested in is $f(0) = a + c$. The HO energy is $\hbar\Omega = 16\,$MeV

Table 3.4 The table shows the extrapolated matter radius, $\overline{r}_{\text{rms}}$, for each truncated $N_{\max}$ space

Ref state	r_{rms} (fm)	$\overline{r}_{\text{rms}}$ (fm)	$\overline{r}_{\text{rms}}$ (fm)	$\overline{r}_{\text{rms}}$ (fm)	$\overline{r}_{\text{rms}}$ (fm)
gs	2.088	2.117	2.142	2.164	2.188
mul	2.088	2.119	2.146	2.186	2.324
Exact	2.088	2.115	2.142	2.168	–

The uncertainty due to the extrapolation procedure is about 10^{-3} fm. In the last row we show the NCSM results for when the $N_{\max}$ basis is complete

an exponential function, given as $f(\kappa) = ae^{-\kappa b} + c$. The extrapolation to $\kappa = 0$ corresponds to determining $f(0) = a + c$.

Since we are extrapolating to $\kappa = 0$, we need to determine what the error from the extrapolating procedure is. As a first attempt to provide some measure of the uncertainty of the proposed extrapolating procedure, we will perform the same extrapolations as shown in Fig. 3.17, in which we will vary the number of data-points used in the extrapolations. The same extrapolating function will be used throughout. Three extrapolations will be performed: (1) use all 12 points as is done in Fig. 3.17, (2) do not include the smallest value of κ, (3) do not include the two smallest values of κ. In other words, we are testing the sensitivity of the extrapolation when we include more points as κ decreases. The central value is determined from the average of the three fits ($\overline{r}_{\text{rms}}$) and the uncertainty ($\sigma_{\text{rms}}$) is determined by determining the spread between the minimum and maximum value of $f(0)$, and dividing it by two.

$$\begin{aligned}\overline{r}_{\text{rms}} &= \frac{\sqrt{\langle r_m^2(\kappa = 0)_{12}\rangle} + \sqrt{\langle r_m^2(\kappa = 0)_{11}\rangle} + \sqrt{\langle r_m^2(\kappa = 0)_{10}\rangle}}{3} \\ \sigma_{\text{rms}} &= \frac{f(0)_{\max} - f(0)_{\min}}{2}\end{aligned} \tag{3.14}$$

We apply our error analysis for each truncated $N_{\max}$ space, and determine the average extrapolated value for the matter radius, as well as the uncertainty from the extrapolations. These results are presented in Table 3.4, along with the NCSM results for a complete $N_{\max}$ basis. Note that the uncertainties are quoted on the level of attometers (a thousandth of a fermimeter), indicating that the extrapolation uncertainties are quite small, in comparison to the actual size of the nucleus. The IT-NCSM calculations done with multiple reference states always predicts a larger RMS radius and overestimate the size of the nucleus as can be seen in Table 3.4.

We can see from Table 3.4 that as $N_{\max}$ increases, so does the size of the nucleus. As we have mentioned before, at $N_{\max} = 14$ the calculation has not converged yet; we need to extrapolate to $N_{\max} = \infty$.

In Fig. 3.18 we show the extrapolation of the RMS matter radius to $N_{\max} = \infty$. The extrapolation is performed (once again) with the function $g(N_{\max}) = a\exp(-b * N_{\max}) + r_{\text{rms},\infty}$, in which a now assumes negative values and $r_{\text{rms},\infty}$ is the RMS matter radius at $N_{\max} = \infty$. The fit is performed taking into account the radii at $N_{\max} = 2$–14. Recall that at $N_{\max} = 2, 4$, we have a complete $N_{\max}$ basis space. For

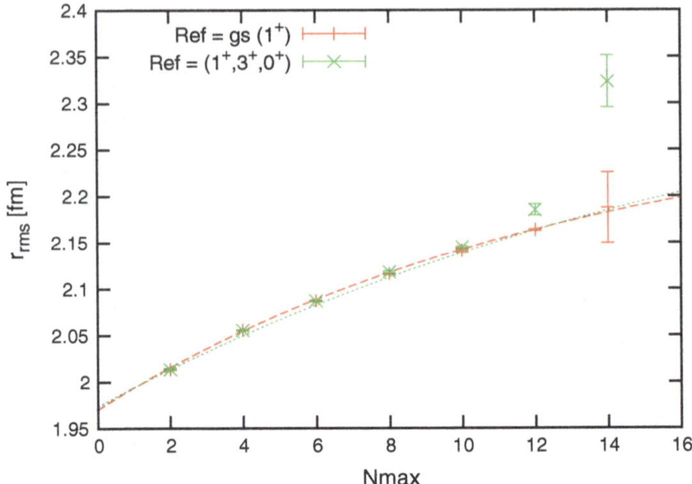

Fig. 3.18 The figure shows the RMS radius for $N_{\max} = 2$–14, which are extrapolated to $N_{\max} = \infty$, by taking into account the numerical and extrapolation uncertainties. The extrapolation is performed using the function $g(N_{\max}) = a\exp(-b * N_{\max}) + r_{\text{rms},\infty}$, in which $r_{\text{rms},\infty}$ is the RMS matter radius at $N_{\max} = \infty$. For the gs reference state, we find that $r_{\text{rms},\infty} = 2.294 \pm 0.015\,\text{fm}$, whereas for the multiple reference states we find $r_{\text{rms},\infty} = 2.383 \pm 0.044\,\text{fm}$

the complete basis spaces, we take the uncertainty to be 1 am, which is the order of numerical uncertainty. We also take into account the uncertainty in the extrapolations to $\kappa = 0$ for the truncated $N_{\max} = 6$–14 spaces. Note that in the case of the reference state being the gs, we find that the $g(N_{\max})$ matches the IT-NCSM results rather well, but in the case of the multiple reference states, we see that extrapolation essentially ignores the points at $N_{\max} = 12$–14. The reason for this is that the uncertainty in the last two points is much larger than the preceding points. Furthermore, another look at Fig. 3.17 suggests that the extrapolation to $\kappa = 0$ in the $N_{\max} = 14$ case (for multiple reference states) is probably largely overestimated. A linear fit to those points might have sufficed, but in order to be consistent with the lower $N_{\max}$ spaces, we retained the exponential extrapolation to $\kappa = 0$. For the gs reference state, we find that $r_{\text{rms},\infty} = 2.294 \pm 0.015\,\text{fm}$, whereas for the multiple reference states we find $r_{\text{rms},\infty} = 2.383 \pm 0.044\,\text{fm}$.

The HO Dependence for r_m^2

The results presented in the previous section had fixed $\hbar\Omega = 16\,\text{MeV}$. We now study how the RMS matter-radius changes as we vary $\hbar\Omega$. Our aim is to determine if IT-NCSM calculations (now for operators) deteriorate in quality as $\hbar\Omega$ increases, as we noticed in the case of the gs energy. In order to judge the quality of the truncated

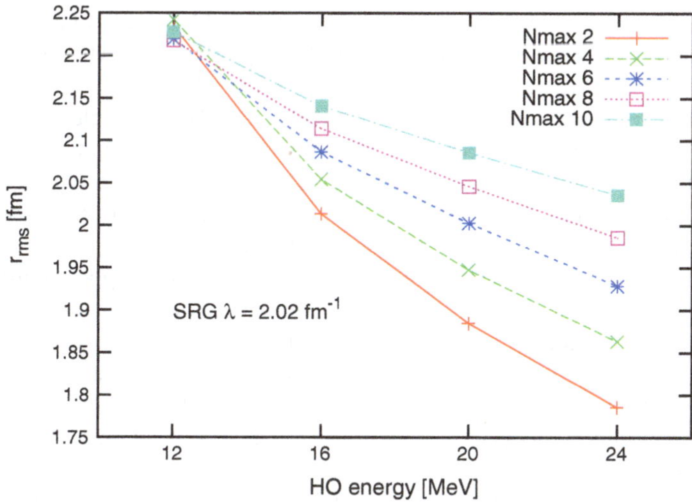

Fig. 3.19 The behavior of r_{rms} in the complete $N_{\max} = 2$–12 spaces as a function of $\hbar\Omega$ (the HO energy). Note that at $\hbar\Omega = 12\,\text{MeV}$, the radius is roughly constant for all values of $N_{\max}$

calculations, we will compare IT-NCSM results with those obtained from a complete $N_{\max}$ basis, which we will refer to as 'exact' calculations.

We would like to remind the reader that there is no variational principle in place for operators other than the Hamiltonian. Thus, the behavior of the matter radius as both a function of $N_{\max}$ and $\hbar\Omega$ might be quite different to that of the gs energy. To illustrate this point, we have plotted the matter radius (r_{rms}) for the complete $N_{\max} = 2$–12 spaces as a function of $\hbar\Omega$ in Fig. 3.19. Note that at $\hbar\Omega = 12\,\text{MeV}$, the radius is roughly constant for all values of $N_{\max}$; this behavior suggests that for calculating the radius, the appropriate $\hbar\Omega$ energy is 12 MeV. Recall that for the gs energy, we preferred to use $\hbar\Omega = 16$–$20\,\text{MeV}$, since the analogous figure to Fig. 3.19 for the gs energy showed a clear minima developing for those ranges of $\hbar\Omega$. At $\hbar\Omega = 12\,\text{MeV}$, the calculations are converging very rapidly, although, in a non-variational way, since the value of r_{rms} oscillates between a range of values as $N_{\max}$ increases. This oscillatory behavior makes extrapolating to $N_{\max} = \infty$ practically impossible.

3.7 Criticisms of the IT-NCSM

The IT-NCSM is still a rather new technique for calculating nuclear-structure observables. Not surprisingly, it has been criticized in various ways [6, 7]. We present the issues of concern here, and try to provide at least some response to the criticism.

Before we begin with any discussion on IT-NCSM calculations, we need to introduce some terminology regarding many-body physics, perturbation theory, and 'exact' techniques. We purposely place the word 'exact' in quotes, as we mean that the calculation is, in principle, exact, up to the 'exactness' of the nuclear Hamiltonian. In other words, provided that one accepts that we have only included NN forces in the Hamiltonian, we can guarantee that the method is exact up to two-body forces. Of course, the omission of the higher-body terms, such as three- or four-body forces, might not reflect the properties of the nucleus in nature.

To introduce these terms, let us remember that we are solving a many-body system, using the NCSM framework (an exact technique). Nuclear matter-calculations, which deal with an infinite system, have traditionally been calculated with the use of perturbation theory; naturally there is at least some approximation in place (e.g., second-order perturbation theory). In the case of IT-NCSM, we are using an *argument* based on perturbation theory, to select a basis by means of the importance measure κ. The truncated basis is the one, in which we 'exactly' diagonalize the Hamiltonian. The combination of perturbation theory and diagonalization in the IT-NCSM has led to some confusion in the field—we aim to clarify the situation somewhat.

3.7.1 Configuration Interaction and NCSM

There are several techniques in use for nuclear-structure calculations that can be considered exact. Most of these exact methods start by expanding the gs wavefunction of the system (Ψ) in a suitable basis. The expansion coefficients of the chosen basis are then determined by variational means, for example, by diagonalizing the Hamiltonian in the chosen basis. But, in what basis should one choose to work? There are two issues to consider: (1) what functional form of the basis should be chosen (e.g., HO basis), and (2) how do the A nucleons fill the single-particle states (i.e., how is the basis truncated)? In the case of this thesis, we only use the HO basis. The second question posed, however, needs to be discussed.

We will now discuss one particular method known as configuration interaction (CI). Configuration-interaction is quite simple to formulate and is also closely related to NCSM ideas; thus, we will explain carefully the key concepts of the method.

To illustrate the concepts of CI calculations, let us assume that we are working with the doubly-magic nucleus ^{4}He. Recall that this is a closed-shell nucleus, and, hence, is described by exactly one Slater determinant (Φ_0), provided we restrict the basis space to only the single-particle HO states corresponding to the $N = 0$ major shell. The desired wavefunction (Ψ) is expanded in terms of particle-hole operators ($\hat{C}_i$) as follows,

$$\begin{aligned} \Psi &= \sum_i \hat{C}_i \Phi_0 \\ &= \hat{C}_0 \Phi_0 + \hat{C}_1 \Phi_0 + \hat{C}_2 \Phi_0 + \cdots , \end{aligned} \tag{3.15}$$

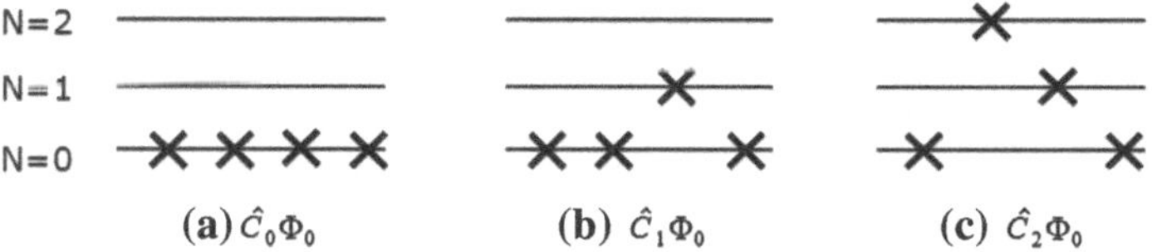

Fig. 3.20 The figure shows the action of the operators $\hat{C_1}$ and $\hat{C_2}$ on the initial Slater determinant Φ_0. Note that $\hat{C_1}$ promotes one of the nucleons into a new single-particle state. The operator $\hat{C_2}$ corresponds to a 2p-2h excitation of Φ_0

in which the omitted terms represent higher-order particle-hole excitations. The action of $\hat{C_i}$ is to generate an i-particle-hole (ip-h) excitation in the reference Slater determinant (Φ_0), in which a particle is removed (creating a hole) from a single-particle state in Φ_0 and is recreated in a new single-particle state, outside the original single-particle configuration of Φ_0. In the case of ^{4}He, the action of the various operators can be seen in Fig. 3.20. Note that we show only one of the many 1p-1h and 2p-2h excitations that are possible.

Equation (3.15) is a short-hand way of expressing the fact that we include up to all 1p-1h and 2p-2h excitations (basis states) in the final diagonalization. In a strict mathematical formulation, the trial wavefunction we displayed in Eq. (3.15) can be written as,

$$\Psi = \Phi_0 + \sum_{i,a}^{A,D} c_i^a \Phi_i^a + \sum_{i>j,a>b}^{A,A-1} c_{ij}^{ab} \Phi_{ij}^{ab} + \cdots \tag{3.16}$$

The subscripts i, j label the single-particle states present in Φ_0, whereas the subscripts a, b label the single-particle states outside the basis space of Φ_0 (i.e., all the single-particle states in the $N = 1$ and $N = 2$ shell). The c_i^a and c_{ij}^{ab} represent the unknown expansion coefficients of the trial wavefunction; these are determined by the diagonalization procedure. Furthermore, note that the index i can assume the values $i = 1 \ldots A$ and that there are D single-particle states present in the final calculation.

We now turn our attention to the inherent truncations that must be made in CI calculations. We will discuss two truncations: (1) the truncation of the basis in terms of how many HO shells are included in the calculation, and (2) restrictions on how the A nucleons fill the single-particle states. In our example, we have already implicitly truncated our single-particle basis. In Fig. 3.20, we truncated the basis at the $N = 2$ HO shell. However, we have not specified how the A nucleons fill the single-particle states. This consideration is important for the remainder of this section.

If we consider Eq. (3.15) for ^{4}He, we note that we can have terms up to 4p-4h excitations ($\hat{C_4}$). If one includes all Ap-Ah excitations in a CI calculation, one speaks of *full CI*. Needless to say, as A increases, the computational cost of including all Ap-Ah excitations becomes unmanageable. Thus, one can start to truncate the particle-hole expansion of the trial wavefunction up to a given order. For example, if we were to truncate Eq. (3.15) at the 2p-2h level, we would speak of a CI

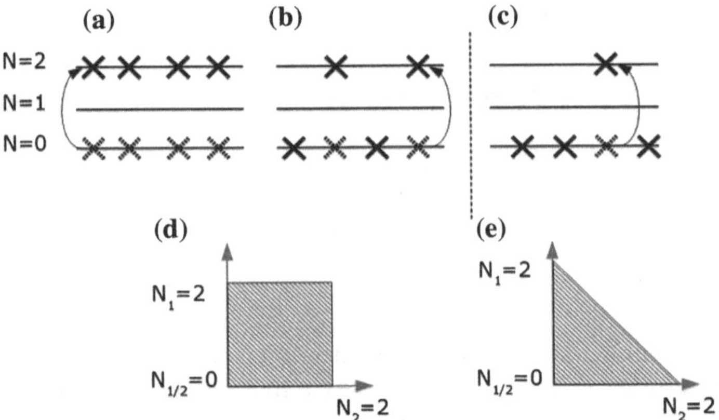

Fig. 3.21 **a** The A nucleons can fill the single-particle basis up to $N = 2$ in full CI. **b** The highest occupied state in CI-SD. **c** The highest occupied state in NCSM for $N_{\max} = 2$. **d, e** The basis truncation is conveniently expressed in terms of a rectangle for CI and a triangle for NCSM calculations. The *shaded area* represents all the possible configurations that are included in the basis. N_1 and N_2 label the HO shell of the first and second nucleons, respectively. Note that in CI calculations, we can place all A nucleons in the highest HO shell, whereas in the NCSM, we can place at most one nucleon in the highest HO shell

calculation that includes (S)ingle and (D)ouble excitations, or CI-SD for short. As an example of how the highest possible single-particle states fill in these two cases, we refer to Fig. 3.21 (see parts a and b) for more detail.

But what does CI have to do with the NCSM? In this regard, the NCSM is a full CI calculation. However, there is a subtle difference in how the basis is constructed. In CI calculations, the basis is truncated on a single-particle level, whereas in the NCSM it is truncated on an energy-quanta level (i.e., N vs. $N_{\max}$). Thus, if we speak of a ^{4}He CI calculation, in which the single-particle basis was truncated at $N = 2$, we do not recover the $N_{\max} = 2$ NCSM basis. To illustrate this point, see parts a and c in Fig. 3.21, in which we have illustrated how the highest single-particle states are occupied in the NCSM at $N_{\max} = 2$. To remind the reader, we state that the NCSM basis is the only basis, in which we can exactly separate spurious center-of-mass motion from the intrinsic states. As a final comment, it is often convenient to express the truncation of CI and NCSM basis spaces in terms of a 'rectangle' and 'triangle' truncation (see diagrams d and e in Fig. 3.21).

3.7.2 Size-Extensivity

Full configuration interaction methods, like the one we described, are usually the best methods for calculating a finite many-body system, because these are variational. However, as we pointed out, including all particle-hole excitations is computationally demanding. We have already described CI-SD, in which we approximate the full CI calculation by truncating the number of particle-hole excitations we include. Another technique which approximates full CI is that of the coupled cluster method [9, 27]. Coupled cluster methods also have the advantage that the calculated energies scale properly with system size, (i.e., linearly in particle number). This property is known as *size-extensivity*. In most many body calculations, the idea of size-extensivity, as well as the closely related property of *size-consistency* is often overlooked. These two ideas are related, and so, to avoid confusion, we will briefly describe them.

Size-consistency is most easily understood from the viewpoint of a chemical reaction. The dissociation of a single system, "AB", into two parts, system A and B, which eventually are infinitely separated from each other, should give the same total energy as the original system "AB". In other words, $E_{AB} = E_A + E_B$. To mathematically define size-consistency is very difficult, and arguments as to whether quantum systems are ever truly separated arise [28].

Size-extensivity on the other hand, states that the energy of a system, such as an electron gas, should scale linearly with the number of particles present. Such an idea is much easier to formulate mathematically. In terms of approximation methods used in many-body techniques, size-extensivity is discussed much more frequently, and is often the most desired property, whereas size consistency is usually an afterthought. Full CI calculations *are* size extensive, since all possible excitations are included. However, truncated CI calculations, like CI-SD, lack size-extensivity. We will illustrate with a simple example. Consider a Helium atom, whose electronic structure is of interest to us. Since the Helium atom has two electrons, we can do a 'full' CI-calculation by calculating the gs energy according to CI-SD. Now consider two non-interacting Helium atoms (4 electrons in total). If we now use CI-SD in calculating the gs energy of the composite system, we lose size-extensivity. The energy of the composite system is now no longer the sum of the individual sub-systems. In order to recover size-extensivity, we would have to include 4p-4h excitations in the composite system (which is the full CI case).

The loss of size-extensivity in infinite systems was first noticed by Brueckner [29]. In the Brueckner calculations, Raleigh-Schroedinger perturbation theory was used to calculate the binding energy per nucleon for the case of nuclear matter. The energy should be linear in N, but terms arising from the Raleigh-Schroedinger perturbation theory (RSPT) expansion contained terms that were proportional to N^2 and N^3. Brueckner showed that these unphysical terms are cancelled up to fourth-order. Goldstone demonstrated that these unphysical terms cancel to all orders [8], since the perturbation terms can be grouped into linked and unlinked diagrams. The unlinked diagrams are the terms that destroy size-extensivity, but, provided one does RSPT to infinite-order, are always cancelled out. The cancellation of unlinked diagrams is known as the Goldstone linked diagram theorem.

CI calculations are related to RSPT expansions, in order to extract the CI eigenvalues. In the case of CI-SD, the calculation retains these unlinked diagrams that are proportional to N^2, etc., and are, thus, not cancelled out. The cancellation occurs if one were to add more excitations to the CI calculation, such as CI with singles, doubles,triples and quadruples included. Unfortunately, unlinked diagrams will still remain, since they are cancelled by an ever higher-order of excitation. The complete cancellation of unlinked diagrams only occurs once all excitations are included, but this brings us back to the computational problems of full CI.

Size-extensivity can be restored in the RSPT expansion. The size-extensivity is restored, if one considers all configurations to a given order. This leads to many-body perturbation theory (MBPT), which is a fully-linked diagrammatic expansion, order-by-order, and is size-extensive, up to that given order [9, 30]. Unfortunately, each subsequent order is more difficult to calculate than the previous one, rendering the method useful, but not efficient.

Coupled cluster (CC) offers a slightly different approach to the problem, by providing an infinite-order resummation of MBPT in selected clusters, such as single and double excitations. CC is, by construction, size-extensive. CI and CC both have the same inherent ideas, i.e., generate single and double excitations on top of a reference state, in a basis truncated on the single-particle level (the rectangle truncation of Fig. 3.21). The difference to CI comes from the exponential ansatz made in CC. The coupled-cluster wavefunction is generated by $\Psi_{\mathrm{CC}} = exp(\hat{T})\Psi_0$, where $\hat{T} = \hat{T}_1 + \hat{T}_2 + \cdots + \hat{T}_n$, where $\hat{T}_p$ is a connected cluster operator that generates the p-fold excitation (similar to the operator $\hat{C}_p$ that we used in the CI discussion). It is this exponential form that ensures the size-extensivity of the method.

In the case of single and double excitations, one speaks of CCSD calculations. The CCSD trial wavefunction assumes the following form,

$$\begin{aligned} \Psi_{\mathrm{CCSD}} &= \exp(1 + \hat{T}_1 + \hat{T}_2)\Phi_0 \\ &= \Phi_0 + (\hat{T}_1 + \hat{T}_2)\Phi_0 + (\hat{T}_1^{\,2} + 2\hat{T}_1\hat{T}_2 + \hat{T}_2^{\,2})\Phi_0. \end{aligned} \tag{3.17}$$

Note that the CCSD trial wavefunction contains the extra non-linear terms such as $\hat{T}_1^{\,2}$, which the CI-SD trial wavefunction does not contain. It can be shown that the $\hat{C}$ operators are related to the $\hat{T}$ operators as follows,

$$\begin{aligned} \hat{C}_1 &= \hat{T}_1 \\ \hat{C}_2 &= \hat{T}_2 + \frac{1}{2}\hat{T}_1^{\,2} \end{aligned} \tag{3.18}$$

We, thus, see that the non-linear terms, such as $\hat{T}_1^{\,2}$, restore the size-extensivity property in CCSD calculations, which are lacking in the CI-SD calculations.

Although CC has some advantages over truncated CI, such as being size extensive and more efficient, it is not as versatile as truncated CI. In the case of nuclear-structure, CC is often used to calculate the gs energies of doubly-magic nuclei, such as ^{40}Ca [31]. Recently, CC in nuclear-structure has been extended to $A \pm 2$ nuclei, in which A represents a doubly magic nucleus [32].

3.7.3 IT-NCSM and Size-Extensivity

Now that we have presented a detailed discussion of the size-extensivity issue, in many-body calculations, we are able to address the issue of size-extensivity in IT-NCSM. This issue was first raised by Dean et al. in a comment that appeared in [6]; a response was given by Roth and Navrátil in [7]. The criticism raised centers around the idea of particle-hole truncations and size-extensivity. A summarize is now presented.

In Sect. 3.3.1 we presented the history of IT-NCSM calculations, concentrating on some of the recent developments. In the original 2007 PRL by Roth and Navrátil [2], the iterative IT-NCSM formulation was used. Recall that the iterative IT-NCSM is based on the idea of a particle-hole truncation scheme. In the case of the ^{4}He calculations, all particle-hole excitations up to 4p-4h were evaluated in the IT-NCSM; thus, it can be considered a full CI calculation, once the extrapolation to $\kappa = 0$ is made. However, Dean et al. argued that in the case of ^{16}O and ^{40}Ca, the restriction to 4p-4h excitations leads to lack of size-extensivity, if $N_{\text{max}} > 4$. In order to return to a full CI picture, one would have to accommodate up to N_{max} particle-hole excitations. Thus, even though the ^{4}He calculation looks very good, one cannot claim that the analogous calculation in ^{40}Ca would be of a similar quality.

The issues raised by Dean et al. are worthwhile to consider, especially in the context of size-extensivity in the iterative IT-NCSM scheme. But what about the sequential IT-NCSM scheme?

Sequential IT-NCSM and Size-Extensivity

The sequential IT-NCSM differs from the iterative IT-NCSM, since we now automatically generate all basis states in the evaluated N_{max} space. Each basis state is evaluated and is kept (or discarded) according to the importance threshold κ_ν. Furthermore, we can generate a sequence of gs energies for each truncated N_{max} basis space, from which we can easily extrapolate the gs energy to $N_{\text{max}} = \infty$ (thus demonstrating convergence as the basis approaches infinity). Finally, there are *no* particle-hole truncations; all possible particle-hole excitations are considered. But what about size-extensivity?

To answer the question regarding size-extensivity, let us restate how the property of size-extensivity is phrased. A size-extensive calculation scales linearly in particle number. In our example, we used the case of two *non-interacting* Helium atoms. But, what do we make of this statement in the case of an *interacting* system? Consider the following: If we calculate the binding energy of ^{16}O in the sequential IT-NCSM (recall all particle-hole excitations are included), should we expect the binding energy of ^{16}O to be four times that of ^{4}He? No, we should not. The ^{16}O nucleus is not the sum of four non-interacting ^{4}He nuclei!

Now, it is true that we have truncated *something*. Indeed, we have excluded many of the basis states present in a given N_{max} space. But, as calculations by Roth have shown [3, 4], as well as those that have been presented in this chapter, once the extrapolation to $\kappa = 0$ has been performed, we generally recover the result of the NCSM $N_{\max}$ calculation, within a few to a hundred keV (approximately), depending on the $N_{\max}$ space. Unfortunately, no clear numerical evidence has been presented up to now that would ultimately rule out the possibility that IT-NCSM does suffer from size-extensive problems. In part, this is so, because no one has posed an unambiguous situation, in which IT-NCSM could be tested for size-extensivity issues. We hope to address this issue in a concrete manner in the future.

References

1. P. Navrátil, S. Quaglioni, I. Stetcu, and B.R. Barrett, Recent developments in no-core shell-model calculations. J. Phys. G: Nucl. Part. Phys. **36**(8), 083101 (2009)
2. R. Roth, P. Navrátil, Ab Initio study of ^{40}Ca with an importance-truncated no-core shell model. Phys. Rev. Lett. **99**, 092501 (2007)
3. R. Roth, Importance truncation for large-scale configuration interaction approaches. Phys. Rev. C **79**, 064324 (2009)
4. P. Navrátil, R. Roth, S. Quaglioni, Ab initio many-body calculations of nucleon scattering on ^{4}He, ^{7}Li, ^{7}Be, ^{12}C, and ^{16}O. Phys. Rev. C **82**, 034609 (2010)
5. R. Roth, J. Langhammer, A. Calci, S. Binder, P. Navrátil, Similarity-transformed chiral $nn+3n$ interactions for the Ab Initio description of ^{12}**C** and ^{16}**O**. Phys. Rev. Lett. **107**, 072501 (2011)
6. D.J. Dean, G. Hagen, M. Hjorth-Jensen, T. Papenbrock, A. Schwenk, Comment on "Ab initio study of ^{40}Ca with an importance-truncated no-core shell model". Phys. Rev. Lett. **101**, 119201 (2008)
7. R. Roth, P. Navrátil, Roth and Navrátil reply. Phys. Rev. Lett. **101**, 119202 (2008)
8. J. Goldstone, Derivation of the brueckner many-body theory. Proc. R. Soc. Lond. Ser. A. Math. Phys. Sci. **239**(1217), 267–279 (1957)
9. J.R. Bartlett, M. Musiał, Coupled-cluster theory in quantum chemistry. Rev. Mod. Phys. **79**, 291–352 (2007)
10. C.D. Sherrill, H.F. Schaefer III, The configuration interaction method: advances in highly correlated approaches, in *Advances in Quantum Chemistry, volume 34 of Advances in Quantum Chemistry*, ed. by M.C. Zerner Per-Olov Lowdin, J.R. Sabin, E. Brandas (Academic Press, New York, 1999), pp. 143–269
11. P.R. Surján, Z. Rolik, A. Szabados, D. Köhalmi, Partitioning in multiconfiguration perturbation theory. Annalen der Physik **13**(4), 223–231 (2004)
12. Z. Rolik, Á. Szabados, P.R. Surján, On the perturbation of multiconfiguration wave functions. J. Chem. Phys. **119**(4), 1922–1928 (2003)
13. P. Navrátil. No core slater determinant code (unpublished, 1995)
14. E. Caurier, F. Nowacki, Present status of shell model techniques. Acta Phys. Pol. B **30**(3), 705 (1999)
15. R. Roth, H. Hergert, P. Papakonstantinou, T. Neff, H. Feldmeier, Matrix elements and few-body calculations within the unitary correlation operator method. Phys. Rev. C **72**, 034002 (2005)
16. D.R. Entem, R. Machleidt, Accurate charge-dependent nucleon-nucleon potential at fourth order of chiral perturbation theory. Phys. Rev. C **68**, 041001 (2003)
17. P. Navrátil, E. Caurier, Nuclear structure with accurate chiral perturbation theory nucleon-nucleon potential: application to ^{6}Li and ^{10}B. Phys. Rev. C **69**, 014311 (2004)
18. P. Navrátil, S. Quaglioni, Ab initio many-body calculations of deuteron-^{4}He scattering and ^{6}Li states. Phys. Rev. C **83**, 044609 (2011)
19. E.D. Jurgenson, P. Navrátil, R.J. Furnstahl, Evolving nuclear many-body forces with the similarity renormalization group. Phys. Rev. C **83**, 034301 (2011)

20. R. Roth, J.R. Gour, P. Piecuch, Center-of-mass problem in truncated configuration interaction and coupled-cluster calculations. Phys. Lett. B **679**(4), 334–339 (2009)
21. M. Thoresen, P. Navrátil, B.R. Barrett, Comparison of techniques for computing shell-model effective operators. Phys. Rev. C **57**, 3108–3118 (1998)
22. E. Caurier, P. Navrátil, Proton radii of ^{4,6,8}He isotopes from high-precision nucleon-nucleon interactions. Phys. Rev. C **73**, 021302 (2006)
23. A.F. Lisetskiy, M.K.G. Kruse, B.R. Barrett, P. Navratil, I. Stetcu, J.P. Vary, Effective operators from exact many-body renormalization. Phys. Rev. C **80**, 024315 (2009)
24. C. Forssén, E. Caurier, P. Navrátil, Charge radii and electromagnetic moments of li and be isotopes from the ab initio no-core shell model. Phys. Rev. C **79**, 021303 (2009)
25. C. Cockrell, J.P. Vary, P. Maris, Lithium isotopes within the ab intio no-core full configuration approach. Phys. Rev. C **86**, 034325 (2012)
26. W.N. Polyzou, Nucleon-nucleon interactions and observables. Phys. Rev. C **58**, 91–95 (1998)
27. G. Hagen, T. Papenbrock, D.J. Dean, M. Hjorth-Jensen, Medium-mass nuclei from chiral nucleon-nucleon interactions. Phys. Rev. Lett. **101**, 092502 (2008)
28. W. Duch, G.H.F. Diercksen, Size-extensivity corrections in configuration interaction methods. J. Chem. Phys. **101**(4), 3018–3030 (1994)
29. K.A. Brueckner, Two-body forces and nuclear saturation. iii. details of the structure of the nucleus. Phys. Rev. **97**, 1353–1366 (1955)
30. R.J. Bartlett, G.D. Purvis, Many-body perturbation theory, coupled-pair many-electron theory, and the importance of quadruple excitations for the correlation problem. Int. J. Quantum Chem. **14**(5), 561–581 (1978)
31. G. Hagen, D.J. Dean, M. Hjorth-Jensen, T. Papenbrock, A. Schwenk, Benchmark calculations for ^{3}H, ^{4}He, ^{16}O, and ^{40}Ca with ab initio coupled-cluster theory. Phys. Rev. C **76**, 044305 (2007)
32. G.R. Jansen, M. Hjorth-Jensen, G. Hagen, T. Papenbrock, Toward open-shell nuclei with coupled-cluster theory. Phys. Rev. C **83**, 054306 (2011)

Chapter 4
UV and IR Properties of the NCSM

4.1 Traditional N_{max} Extrapolations

In this chapter we discuss some of the formal aspects of NCSM calculations and the subsequent extrapolations to $N_{max} \to \infty$. The NCSM calculations, even when the bare Hamiltonian is used, depend on two parameters: N_{max} (the size of the basis) and $\hbar\Omega$ (the HO 'frequency' chosen for the basis). To illustrate this dependence, we show the Triton gs energy as a function of $N_{max} = 8 - 16$ and $\hbar\Omega$ in Fig. 4.1, in which we used the chiral NN N3LO interaction, which has been regulated at 500 MeV/c [1]. A plot, such as Fig. 4.1, is referred to as an $(N_{max}, \hbar\Omega)$ mesh-plot.

Figure 4.1 shows the variational nature of the NCSM, when the bare Hamiltonian is used. There are some interesting features to note. For one, note that the gs energy decreases as N_{max} increases, and also seems to display the onset of some converged behavior for some values of $\hbar\Omega$. Furthermore, as N_{max} increases, we notice that the dependence on $\hbar\Omega$ seems to weaken around a range of $\hbar\Omega$ values, located near the minimum of each fixed N_{max} curve. It is not too hard to imagine that as N_{max} approaches infinity, that the $\hbar\Omega$ dependence will disappear completely.

For $N_{max} \to \infty$ extrapolation purposes, the general procedure is as follows. First, one locates the variational minimum in the largest N_{max} space that was performed, in our case $N_{max} = 16$. For Fig. 4.1, the variational minimum is located at $\hbar\Omega =$ 30 MeV. Next, a series of N_{max} gs calculations are performed, in which $\hbar\Omega$ is now fixed at $\hbar\Omega = 30$ MeV. These gs energies are extrapolated to $N_{max} = \infty$ as a function of N_{max}, using the functional form $E_{gs}(N_{max}) = a * \exp(-b * N_{max}) + E_{gs}(\infty)$. The aim is to determine $E_{gs}(\infty)$.

The extrapolation procedure we have described is the one we have commonly used in this thesis and is commonly used by others [2, 3]. We should point out that there are refinements that can be made to the procedure, in which one can determine an error from the extrapolation technique [4]. Providing error estimates in NCSM calculations has been a recent inclusion, in part to provide reasonable uncertainty quantification of the NCSM methods. However, we will not discuss the general error estimates here.

M. K. G. Kruse, *Extensions to the No-Core Shell Model*,
Springer Theses, DOI: 10.1007/978-3-319-01393-0_4,
© Springer International Publishing Switzerland 2013

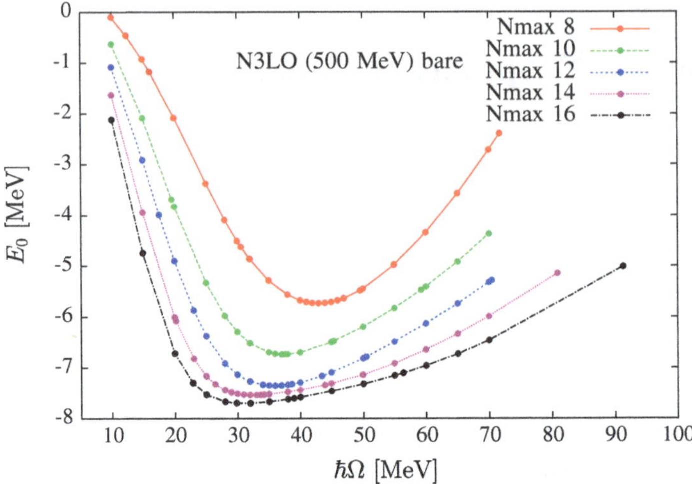

Fig. 4.1 The gs energy of the Triton as a function of $N_{\max}$ and $\hbar\Omega$. The NN interaction employed is the bare chiral N3LO interaction, which has been regulated at 500 MeV

We will instead focus our attention on what physical meaning the extrapolation may contain. In this regard, our main aim is to address the extrapolation of the gs energy, into a regime, where we believe the energy is free of the NCSM parameters. Historically, it has been argued that extrapolating to $N_{\max} = \infty$ 'frees' the gs energy from any residual dependence on $N_{\max}$ and $\hbar\Omega$. However, this argument is based on evidence that 'it has worked in the past'. Indeed, we can see from the mesh-plot that as $N_{\max}$ increases, that (a) the relative change in the gs energy decreases, (b) the $\hbar\Omega$ dependence weakens and (c) the variational minimum slowly shifts to a lower $\hbar\Omega$ value. A purist might argue, 'but how do you know that this trend continues as $N_{\max}$ increases?'. Furthermore, since the variational limit tends to shift to lower values as $N_{\max}$ increases, does the trend perhaps imply that $\hbar\Omega \to 0$ in the infinite basis space?

Taking a step back, most certainly everyone will agree that $N_{\max} \to \infty$ is one of the correct limits to consider when extrapolating to the infinite basis space. However, the other parameter, $\hbar\Omega$ is often ignored. The No-Core Full-Configuration (NCFC) [4], a variant on the extrapolation of the NCSM, has tried to improve on the situation. In the NCFC, multiple extrapolations to $N_{\max} = \infty$ are done at various values of $\hbar\Omega$; furthermore, a reliable error estimate is supplied with each extrapolation, which is a great improvement over the standard NCSM extrapolation technique. However, one *still* treats the $N_{\max}$ parameter separate from $\hbar\Omega$, in effect, by stating that $N_{\max} \to \infty$ is the only limit one needs to consider.

Are we justified to treat these two parameters on an unequal footing? That is the overall theme of this chapter.

4.2 Reformulating NCSM Parameters into an EFT Language

We made the suggestion at the end of the introduction (Sect. 4.1) that the NCSM parameters $N_{\max}$ and $\hbar\Omega$ should be treated on an equal footing. In order to do that, we need to rely on some concepts of Effective Field theory (EFT). In an EFT setting, one has control of the errors in the calculation, and can provide a systematic improvement on the calculation as one increases the model-space size. In this case, we mean 'model-space' to be a general space, which is defined to be useful in the EFT setting. Furthermore, we define error in this context to mean that the calculated gs energy in a finite model-space is in 'error', compared to the infinite model-space (i.e., one has to extrapolate the finite model-space calculations to the infinite model-space). Thus, the first thing we would like to do is to reformulate NCSM parameters into an EFT-style model-space, as done in the initial work of [5–8].

The mapping between NCSM model-spaces and EFT model-space is actually simple to do. We do so by introducing two momentum regulators in the EFT model-space, namely, the ultra-violet (UV) and infra-red (IR) regulator. By doing so, we have done two things: (1) we have placed the two scales[1] on an equal footing relative to each other (thus being consistent) and (2) we gain control of the error in the calculation. We can gain control of the error in the calculation, since we implicitly assume that the error tends to zero as the UV (IR) regulator is taken to infinity (zero). In other words, as the two regulators approach the limit of the infinite model-space, the error made in the finite model-space calculations will tend to zero.

4.2.1 The UV and IR Momentum Regulators

The UV momentum regulator is defined to be,

$$\Lambda = \sqrt{m_N(N_{\max} + 3/2)\hbar\Omega}, \tag{4.1}$$

in which $m_N = 938.92\,\text{MeV}$ represents the average nucleon mass. This definition is a simple application of the continuum definition to the discrete HO basis. To 'derive' Λ, one applies the virial theorem to the highest HO level to establish that the kinetic energy is one half of the total energy. One then solves for the momentum (note that factors of 1/2 will cancel). Recall that in the $N_{\max}$ truncation, the highest occupied single-particle state lies at $N_{\max}$ HO quanta above the unperturbed ($N_{\max} = 0$) configuration. In the case of $0s$-shell nuclei, this implies that the highest occupied single-particle state is found in the major HO shell $N = N_{\max}$. Thus, our definition of the UV momentum regulator is a statement that the highest occupied single-particle state defines the maximum momentum of our finite model-space.

[1] A scale is a physical property of the system, in this case, set by the nuclear interaction, whereas a regulator reflects a variable mathematical quantity.

In defining the IR momentum regulator, there are two possibilities. This, in itself, is interesting to the community, since there has been much debate about which definition of the IR regulator is the correct one. We address this issue in the later sections.

The two proposed IR momentum regulators are given by,

$$\lambda = \sqrt{m_N \hbar\Omega} \tag{4.2}$$

$$\lambda_{sc} = \sqrt{\frac{m_N \hbar\Omega}{(N_{\max} + 3/2)}}. \tag{4.3}$$

The λ definition [5] is argued from the viewpoint, that since we are using the HO basis, the single-particle states corresponding to different major HO shells are separated in energy by $\hbar\Omega$. This is, of course, so, because the energy levels in any quantum system are quantized, when there is a finite confining volume. Note, however, that there is *no* external confining HO potential in place. The $\hbar\Omega$ dependence is due to the underlying HO basis. The limit $\lambda \to 0$ makes physical sense, since we are then removing the artificial IR momentum dependence on the system.

The λ_{sc} definition can be derived from considerations of the spatial extent of the single-particle one-dimensional HO functions [9]. One can show that the spatial extent, on which the HO wavefunctions can accurately describe an object, is given by $r^2 = \frac{N\hbar^2}{m_N \hbar\Omega}$. As N increases, the coordinate wavefunctions spread out over a larger spatial extent, allowing us to resolve larger objects (or equivalently smaller momenta). The opposite is true as $\hbar\Omega$ increases; we are unable to describe large objects. In relating Eq. (4.2) to λ_{sc}, we have used the Heisenberg relationship between momentum and position, and have also included in the zero-point motion of the HO (the additional 3/2 term).

4.2.2 Which IR Momentum Regulator is the Correct One?

All our calculations of gs energies are performed in a finite model-space, be it in the $N_{\max}$ and $\hbar\Omega$ $N_{\max}$ model-spaces or the UV and IR EFT model-spaces. This is, of course, so, because we never include an infinite basis in our calculations, although we do try to include as many basis states as possible. In an EFT framework, the proper way to extrapolate to the infinite model-space limit is to let the UV and IR regulators approach the limits of infinity and zero, respectively. This particular line of thought can be understood in Fig. 4.2, in which we show the two momentum regulators (UV and IR) fixed in a schematic finite model-space.

Figure 4.2 displays a particularly convenient feature of the EFT model-space. If we truly believe that this is the correct model-space, then according to the concepts of EFT model-spaces, if we fix the IR and let the UV regulator increase, the relative error of our calculations should become smaller. Conversely, if we fix the UV regulator and lower the IR regulator, we also expect the relative error to decrease. This idea

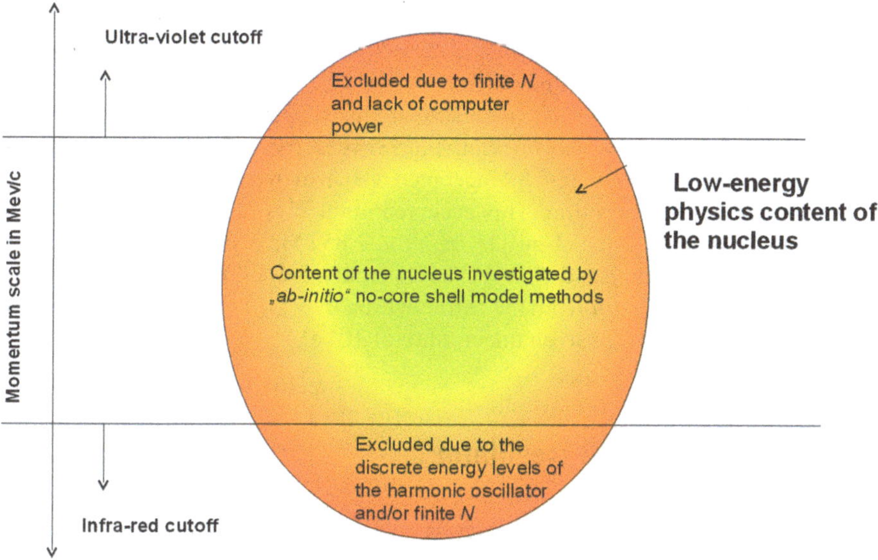

Fig. 4.2 The figure shows a schematic view of a finite model-space, in which the UV and IR momentum regulators are arbitrary. To reach the infinite model-space, one needs to let the UV$\rightarrow \infty$ and the IR $\rightarrow 0$

is one of the key tests we will use in order to determine, which one of the two IR regulators is the 'better' choice in an EFT model-space. To be specific, we define the relative error to be,

$$|\frac{\Delta E}{E}| = |\frac{E(\Lambda, \lambda_{IR}) - E}{E}|, \tag{4.4}$$

in which λ_{IR} denotes either IR momentum regulator (depending on which IR regulator we are investigating) and E is considered to be an exact theoretical result.

To test these ideas, we will use the Chiral NN N3LO interaction, which has the UV divergences regulated at 500 MeV/c [1]. It is important to note that the interaction does not have a sharp cutoff at exactly $\Lambda_{NN} = 500$ MeV/c, since the terms in the interaction are actually regulated by an exponentially suppressed term of the form

$$\exp\left[-\left(\frac{p}{\Lambda_{NN}}\right)^{2n} - \left(\frac{p'}{\Lambda_{NN}}\right)^{2n}\right]. \tag{4.5}$$

In Eq. (4.5), $n \geq 2$ for NLO and higher-order terms and p and p' denote the initial and final momenta in the center-of-mass frame, respectively. Thus, even if we were to raise the UV regulator above 500 MeV/c, we have still not exhausted all the UV physics that is present in the interaction. This interaction is used in the NCSM calculations for a variety of nuclei, such as the Deuteron, Triton and ^{4}He nucleus. These three nuclei are all quite different. The deuteron, for example, is very weakly bound, and, thus, is expected to be sensitive to the IR regulator. ^{4}He, on the other hand, is a closed-shell nucleus and, thus, rather tightly bound, making it less sensitive to IR physics. The Triton is somewhere between these two extreme cases. For that reason, we will mainly present Triton results in this chapter.

In order to determine what our relative error is in the subsequent calculations, we will use what is considered the accepted value of the Triton binding energy for the specified NN interaction. This accepted number is $E = -7.855$ MeV from a 34 channel Faddeev calculation [1], $E = -7.854$ MeV from a hyperspherical harmonics expansion [10], and $E = -7.85(1)$ from a NCSM calculation [2]. In the NCSM calculation, the error from the extrapolation to $N_{\max} = \infty$ is denoted as '(1)' and in this case corresponds to an uncertainty of 10 keV.

4.3 Setting λ as the IR Regulator

In this section we assume that the correct IR regulator is given by λ. We now proceed to perform calculations in the NCSM, in which we first fix the IR regulator (λ) and vary the UV (Λ) regulator. In Fig. 4.3 we show the results of this calculation. It can be seen from Fig. 4.3 that as we increase the UV regulator (Λ), that indeed the relative error $|\frac{\Delta E}{E}|$ decreases, regardless of the value of the IR regulator. Note that the relative error decreases as we lower the IR regulator; it increases as the IR regulator increases. This agrees with our notion that as both momentum regulators approach their respective infinite limits that the relative error tends to zero.

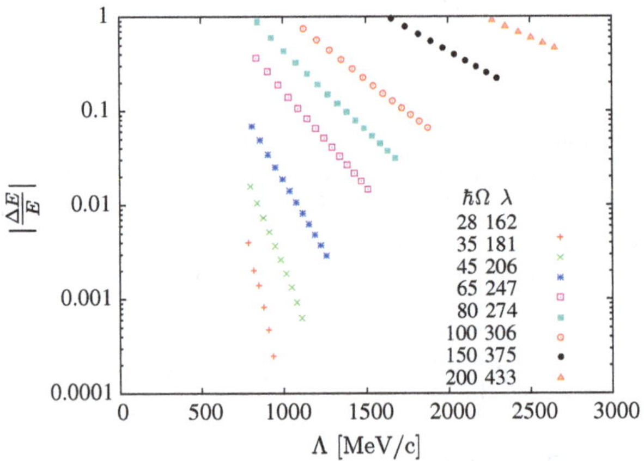

Fig. 4.3 The figure shows the Triton gs energy, as compared to the exact result of $E = -7.85$ MeV. Note that the relative error $|\frac{\Delta E}{E}|$ decreases as we increase the UV (Λ) regulator. Note that we have fixed the IR regulator (λ). The legend shows the respective $\hbar\Omega$ and λ values

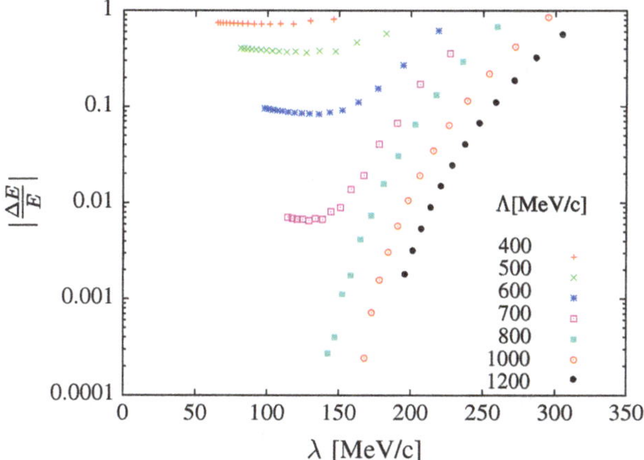

Fig. 4.4 The figure shows the Triton gs energy, as compared to the exact result of $E = -7.85$ MeV. Note that the relative error $|\frac{\Delta E}{E}|$ decreases as we decrease the IR (λ) regulator. However, if $\Lambda < 800$ MeV/c, then the relative errors starts to increase again. In that case, the excluded UV physics dominates the size of the relative error

We now reverse the situation and keep Λ fixed and allow the IR regulator to approach zero ($\lambda \to 0$). In this case, we expect that the relative error should also decrease as the IR regulator is lowered, regardless of the initial UV regulator.

We can see from Fig. 4.4, that the relative error does decrease as we let $\lambda \to 0$. However, it does so only if the UV regulator is above a certain threshold, $\Lambda > 800$ MeV/c. What does this mean physically? It means that one lowers the IR regulator up to a point, but if the UV regulator is not 'large' enough, then the contribution to the relative error starts to be dominated by the UV physics that has been excluded in the calculation. Earlier, we had stated that we should not expect the UV scale to be around 500 MeV/c, as the name of the NN interaction suggests. The nuclear interaction is calculated in relative coordinates, which implies that its effective UV scale is set by the expression $\Lambda^{NN} = \sqrt{\mu(N_{\max} + 3/2)\hbar\Omega}$, in which μ is the reduced mass of the two-nucleon system. In order to identify this scale with our HO UV definition, we note that the reduced mass introduces a factor of $\sqrt{2}$ as compared to the original expression. Thus, one expects the NN interaction to contain UV physics up to about $\sqrt{2}\Lambda^{NN} = 780$ MeV, if one emulates the interaction in the HO basis. This numerical estimate agrees fairly well with what Fig. 4.4 shows; the UV regulator needs to be above 800 MeV/c in order to have the dominant contributions to the relative error be from the excluded IR physics (and not from the excluded UV physics).

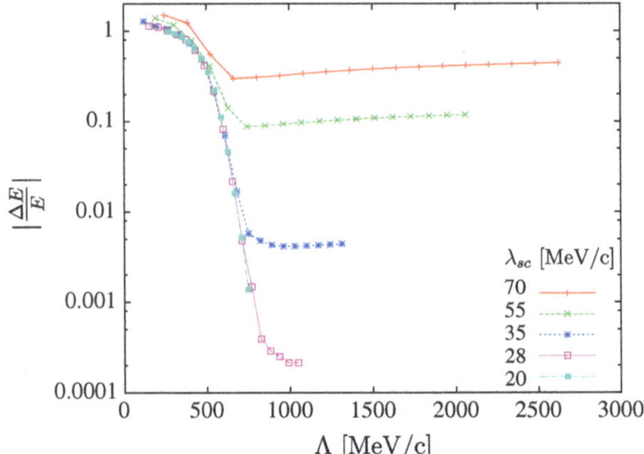

Fig. 4.5 The gs energy of the Triton. The figure shows the relative error as a function of the UV (Λ) regulator at fixed IR values (λ_{sc}). Note that the relative error is smaller if the IR regulator is decreased. The growth in the relative error as the UV regulator increases is not well understood

4.4 Setting λ_{sc} as the IR Regulator

In this section we use the proposed IR regulator of λ_{sc}. To remind the reader, note that λ and λ_{sc} are related by,

$$\lambda_{sc} = \frac{\lambda}{\sqrt{N_{\max} + 3/2}} = \frac{\lambda^2}{\Lambda}. \tag{4.6}$$

This definition states that one actually lowers the IR regulator as the basis approaches completeness ($N_{\max} \to \infty$).

We now test our EFT ideas using this particular IR regulator. Once again, we begin by fixing the IR regulator (λ_{sc}) and increase the UV regulator arbitrarily (see Fig. 4.5).

We see in Fig. 4.5 a number of interesting features. First, note that all curves tend to lie on top of each other for low UV values. Secondly, note that in the region $500 < \Lambda < 1{,}000$ MeV/c, that the relative error decreases up to a point, at which, depending on the value of IR regulator (λ_{sc}), the relative error slowly increases again. In the range of $\Lambda \approx 800$ MeV/c we are including all the UV physics of the NN interaction, which explains the decrease in the size of the relative error. Conversely, the rise in the relative error for $\Lambda > 1{,}000$ MeV/c could be a sign that we have exhausted all the UV physics information contained in the NN interaction, and that this growth might be due to missing IR physics that we have excluded. The curve however tends to increase as the UV regulator increases, instead of remaining constant—this is not well understood at the moment.

It has been suggested by Prof. Furnstahl that this behavior seems to be due to the way that the HO phase space has been truncated. By arbitrarily fixing the IR

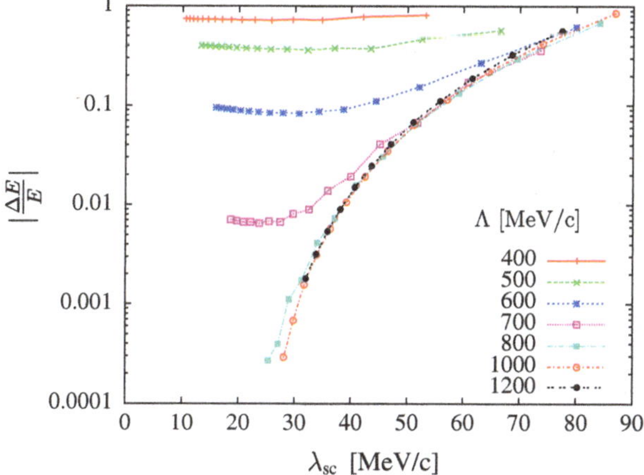

Fig. 4.6 The Triton gs energy. The figure shows the relative error in the gs calculation for fixed UV values (Λ) as a function of the IR regulator, λ_{sc}. Note that above $\Lambda \geq 800$ MeV/c, that all the curves lie on a universal curve

regulator, and letting the UV vary, we have in effect truncated the phase space in an unintended way.[2]

Figure 4.6 shows the analogue case of Fig. 4.4, in which we fix the UV regulator and let the IR regulator vary. The figure shows the relative error in the gs calculation for fixed UV values (Λ) as a function of the IR regulator, λ_{sc}. Note that above $\Lambda \geq 800$ MeV/c all the curves lie on a universal curve. This is the reason why λ_{sc} is often called '*lambda scaling*'. Furthermore, this scale of 800 MeV/c once again seems to suggest that this is the UV scale of the underlying NN interaction. The rise in the relative error is due to missing UV physics that has not been included in the EFT model-space.

4.4.1 Properties of λ_{sc}

The scaling behavior we displayed in Fig. 4.6 is quite interesting. We note that it is not unique to the Triton case, but is also seen in other light nuclei (see Fig. 4.7). Once again, we note that the UV regulator needs to be larger than 800 MeV/c for the scaling behavior to set in. Figure 4.7 also shows quite elegantly that the deuteron is the most sensitive to IR physics (since it is so loosely bound) whereas ^{4}He is the least sensitive (since it is so tightly bound). This can be seen by what values λ_{sc} takes on for these nuclei.

The scaling behavior also seems to suggest a practical application, namely, that we could use it as a means to extrapolate to the infinite EFT model-space. Recall that

---

[2] Private communication at the DNP conference in 2011.

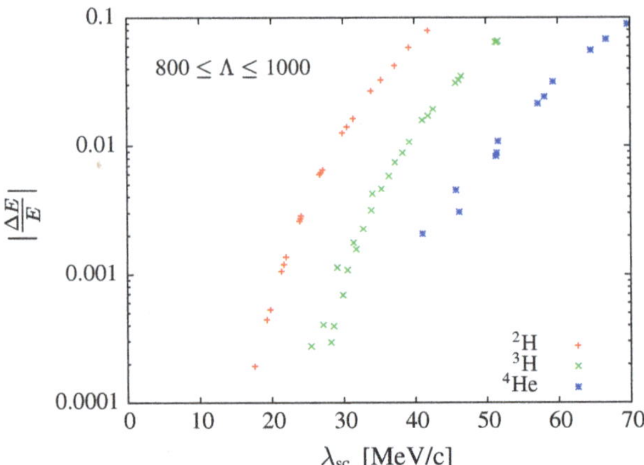

Fig. 4.7 The figure shows the gs energy of three light nuclei: the deuteron, Triton and ^{4}He. Note that once $\Lambda \geq 800$ MeV, the scaling behavior sets in. One can also see that our intuition of the deuteron being the most sensitive to IR physics, and ^{4}He the least sensitive, is correct

we need to capture both the UV and IR physics present in the many-body system. As we have argued several times, the UV scale seems to be set around 800 MeV/c, so provided we obtain gs energies with $\Lambda \geq 800$ MeV/c, we have captured all the UV physics. The IR limit is recovered by taking $\lambda_{sc} \rightarrow 0$. In order to perform the extrapolation, we propose that the scaling curves can be fit by an exponential function, in this case given by $E(\lambda_{sc}) = a * \exp(-\frac{b}{\lambda_{sc}}) + E(\lambda_{sc} = 0)$. We are, of course, interested in determining $E(\lambda_{sc} = 0)$.

In Fig. 4.8, we show the result of extrapolating the gs energy of the Triton to the IR limit $\lambda_{sc} \rightarrow 0$. This is done for several values of Λ, since in principle, if we have captured all the UV physics, any value of the UV regulator should lead to the same gs energy. Indeed, the result of the extrapolations are that $E(\lambda_{sc} = 0) = 7.85 \pm 0.001$ MeV. In other words, we agree with the other theoretical calculations that have been done for the Triton gs energy using the Chiral NN interaction. Furthermore, our error estimate of 1 keV is 10 times smaller than the error estimate in the NCSM extrapolation! This is quite satisfactory.

In Fig. 4.5 we noted that the rise in the relative error increases as the UV regulator is increased. In Fig. 4.9 we demonstrate that this trend is also seen for the other two nuclei, namely the deuteron and ^{4}He. This time we have displayed only one fixed value of the IR regulator, $\lambda_{sc} = 55$ MeV/c. One can also see from the figure that the relative error is larger for the deuteron than it is for the Triton, and is the smallest for ^{4}He. This goes back to the fact that the relative error is larger for systems that are weakly bound.

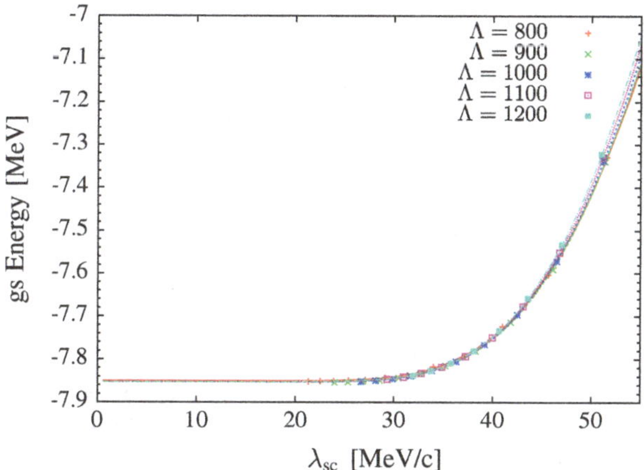

Fig. 4.8 The result of extrapolating the gs energy of the Triton to the IR limit $\lambda_{sc} \rightarrow 0$ for various fixed UV regulators. Taking into account the five extrapolations shown, we determine $E(\lambda_{sc} = 0) = 7.85 \pm 0.001$ MeV, in agreement with other theoretical calculations

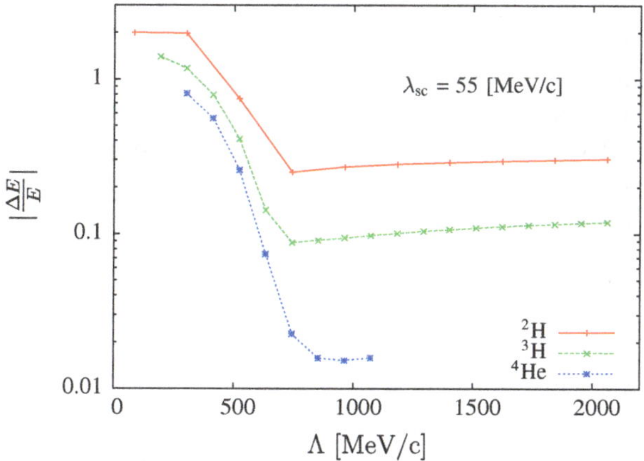

Fig. 4.9 The figure shows the relative error as a function of the UV (Λ) regulator at fixed IR ($\lambda_{sc} = 55$ MeV/c). Note that the rise in the relative error is seen for all three nuclei and is larger for systems that are weakly bound, such as the deuteron

4.5 Conclusions

In this chapter we have reformulated the traditional NCSM model-space of $(N_{\max}, \hbar\Omega)$ into an EFT-style model-space, characterized by UV and IR regulators, (Λ, λ_{IR}). In the EFT model-space, we treat the UV and IR regulators on an equal footing. This is in contrast to the case of the traditional NCSM extrapolations, in which $N_{\max}$ is generally taken to be 'more important' than $\hbar\Omega$. Furthermore, in the EFT model-space, one has complete control over the UV and IR physics in the many-body problem. To this extent, a deficiency in a many-body calculation, as characterized by the size of the relative error $|\frac{\Delta E}{E}|$, if E is known, can be attributed to excluded UV or IR physics. By raising the UV or lowering the IR regulator, we should decrease the size of the relative error automatically.

The UV regulator has generally been accepted to be one of the form $\Lambda \sim \sqrt{N\hbar\Omega}$, in which N represents either the highest lying major HO shell or N_{max}. On the other hand, there has been some debate on what the IR regulator should be. Two possible choices have been proposed: (1) $\lambda \sim \sqrt{\hbar\Omega}$, which is based on an argument that the IR regulator is set by the quantization in energy of the HO shells and (2) $\lambda \sim \sqrt{\frac{\hbar\Omega}{N}}$, which is based on the spatial extent of a single-particle HO (1-D) wavefunction.

In our first study, we tested either IR regulator as being a candidate for the appropriate choice. Our test is based on fixing either momentum regulator in the problem, and taking the other (non-fixed) regulator to the appropriate infinite model-space limit. For example, by fixing the IR regulator, we let the UV regulator extend to infinity. In this procedure, we analyze the behavior of the relative error of the gs energy of the Triton as we vary either momentum regulator. A decrease in the relative error as either regulator is taken to the appropriate limit ($\Lambda \to \infty$ or $\lambda_{IR} \to 0$) is taken as a sign that the candidate IR regulator (λ or λ_{sc}) is 'behaving' like an IR regulator should.

Our calculations provide several insights. When we use the Chiral NN N3LO interaction, which has its UV divergences regulated at 500 MeV/c, we note that we require $\Lambda \geq 800$ MeV/c in order to capture all the UV physics present in the many-body problem. The reader should not be alarmed that the UV scale is higher than the 500 MeV/c we quoted. Recall that the interaction is defined in relative coordinates, which contains an extra factor of $\sqrt{2}$ (due to the presence of the reduced mass) as compared to the HO basis UV definition Λ. Since the interaction is emulated in an HO basis, we should expect that the UV cutoff is approximately at $\sqrt{2} * 500 \approx$ 800 MeV/c.

The choice for the IR regulator leads to some interesting discussion. In the case of λ, we find that the relative error decreases when λ is fixed and Λ is raised; as we expect. However, in the reverse situation, in which we fix Λ and lower λ, we find that the relative error actually increases for $\Lambda \leq 700$ MeV/c. However, we also note that if $\Lambda \geq 800$ MeV/c, we find that the relative error tends to decrease as we lower the IR regulator. The fact that the relative error increases for $\Lambda \leq 700$ MeV/c can be understood from the viewpoint that the errors, arising from excluding UV physics, dominate the size of the relative error over those errors that arise from having the IR regulator too large.

In the case of λ_{sc} we too find some interesting behavior in the relative error. When we let $\Lambda \to \infty$ at fixed λ_{sc}, we note that the relative error increases once $\Lambda \geq 800$ MeV/c, and that it is larger for larger values of λ_{sc}. In this case, it seems that the error made in excluding IR physics is dominating the contribution to the relative error. Recovering the IR limit, in which we fix Λ and let $\lambda_{sc} \to 0$ we find a very interesting set of curves. If the UV regulator is set high enough ($\Lambda \geq 800$ MeV/c), we find that regardless of the value of Λ, all curves lie on the same universal curve and furthermore extrapolate to the same gs energy. In the case of the Triton gs energy, we determine $E(\lambda_{sc} = 0) = -7.85 \pm 0.001$ MeV, in agreement with other theoretical calculations that have employed the same NN interaction.

It is the universal behavior of λ_{sc} that leads us in the direction of considering it to be the appropriate definition of the IR regulator. Provided one captures all the UV physics present in the many-body problem, the extrapolation to $\lambda_{IR} \to 0$ will lead us to the correct gs energy. The previous statement implicitly states that one does not have to take $\Lambda \to \infty$ necessarily; one just needs to have it large enough to capture all the UV physics. Now, in the case of some NN interactions, particularly ones such as Av_{18}, the required scale could be quite a bit higher than the 800 MeV/c that we have been quoting in this chapter. Thus, for other interactions, one will first have to determine the extent of the UV physics in the interaction, before any extrapolations using λ_{sc} are attempted.

As a final comment, the definitions of the UV (Λ) regulator and IR momentum regulator, be it either λ or λ_{sc}, should be considered to be the leading-order terms in a more general definition. Presently, there has been no attempt to derive higher-order terms. Furthermore, some of the peculiar features we have seen, such as the rise in the relative error as one of the momentum regulators approaches the appropriate limit (see Fig. 4.5), could be explained by these higher-order terms.

References

1. D.R. Entem, R. Machleidt, Phys. Rev. C **68**, 041001 (2003). http://link.aps.org/doi/10.1103/PhysRevC.68.041001
2. P. Navrátil, E. Caurier, Phys. Rev. C **69**, 014311 (2004). http://link.aps.org/doi/10.1103/PhysRevC.69.014311
3. C. Forssén, J.P. Vary, E. Caurier, P. Navrátil, Phys. Rev. C **77**, 024301 (2008). http://link.aps.org/doi/10.1103/PhysRevC.77.024301
4. P. Maris, J.P. Vary, A. M. Shirokov, Phys. Rev. C **79**, 014308 (2009). http://link.aps.org/doi/10.1103/PhysRevC.79.014308
5. I. Stetcu, B. Barrett, U. van Kolck, Phys. Lett. B **653**, 358 (2007a). ISSN issn 0370–2693, http://www.sciencedirect.com/science/article/pii/S0370269307009185
6. I. Stetcu, B.R. Barrett, U. van Kolck, J.P. Vary, Phys. Rev. A **76**, 063613 (2007b). http://link.aps.org/doi/10.1103/PhysRevA.76.063613
7. I. Stetcu, J. Rotureau, B. Barrett, U. van Kolck, Ann. Phys. **325**, 1644 (2010). ISSN issn 0003–4916, http://www.sciencedirect.com/science/article/pii/S0003491610000400
8. J. Rotureau, I. Stetcu, B.R. Barrett, M.C. Birse, U. van Kolck, Phys. Rev. A **82**, 032711 (2010). http://link.aps.org/doi/10.1103/PhysRevA.82.032711
9. E.D.Jurgenson (2009), Ph.D. Thesis (Advisor: R.J. Furnstahl). arXiv:0912.2937[nucl-th]
10. A. Kievsky, S. Rosati, M. Viviani, L.E. Marcucci, L. Girlanda, J. Phys. G: Nucl. Part. Phys. **35**, 063101 (2008). http://stacks.iop.org/0954-3899/35/i=6/a=063101

Chapter 5
Extending the NCSM with the RGM

5.1 Introduction

In this chapter we discuss the physics of *exotic* nuclei. These nuclei are located 'far away' from the valley of stability. The valley of stability is so named, since the stable nuclei are energetically stable against $\beta-$decay. Before we turn our attention to these exotic nuclei, we have to paint a general picture, in which each tiny mosaic tile (each individual nuclide), when put together as a whole, displays the rich behavior seen across the nuclear landscape.

5.1.1 *Nuclei Away from Stability*

In Fig. 5.1 we show three isotopic chains of nuclei (He, Li and Be), as they are typically presented in the nuclear chart. We indicate the stability of the nuclei (in gray) or if they are unstable (in white) their half-lives. Let us consider the Lithium chain, as a representative example. We note that there are two stable isotopes, ^{6}Li and ^{7}Li. To the left of the stable isotopes, we find two proton-rich isotopes (both unstable), ^{4}Li and ^{5}Li. The $A = 5$ system, of which ^{5}Li is one of the two isobars that are known, is interesting to study, especially as a test for theoretical nuclear forces. This system can be thought of as an $\alpha-$core with either a neutron (^{5}He) or proton (^{5}Li) orbiting the core. Yet, both nuclei are not stable, and, in fact, have very short half-lives. They do, however, give a stringent test of the spin-orbit force. For the unfamiliar reader, the spin-orbit force is responsible for the splitting of the p-shell into a $p_{3/2}$ and $p_{1/2}$ level. Thus, if one analyzes the phase-shift obtained from a neutron scattering off ^{4}He, one is directly probing the $p_{3/2}$ and $p_{1/2}$ level splitting. Recently, two theoretical calculations were done, in which the phase-shift of the neutron (scattering off ^{4}He) as a function of kinetic energy was determined [1, 2]. When only NN forces were used, the spin-orbit splitting between the $^2P_{3/2}$ and $^2P_{1/2}$ channels was found to be too small. However, with the inclusion of a NNN force,

M. K. G. Kruse, *Extensions to the No-Core Shell Model*,
Springer Theses, DOI: 10.1007/978-3-319-01393-0_5,
© Springer International Publishing Switzerland 2013

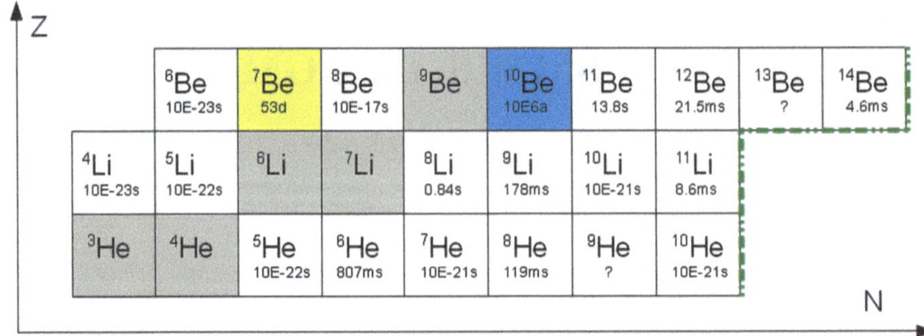

Fig. 5.1 The isotopic chains of He, Li and Be. Gray squares indicate stable nuclei, whereas white squares indicate unstable nuclei. For unstable nuclei, their respective half-lives are indicated. The colored squares indicate nuclei that have a lifetime on the order of days (^{7}Be) or millions of years (^{10}Be). The green dashed line indicates the neutron-drip line

the agreement between experiment and theory drastically improved, to the point that in one calculation the theoretical and experimental results agreed almost uniformly [1]. From a theoretical aspect, it is calculations such as these, that allow us to gain valuable insight into the nuclear force, such as the role of NNN forces in many-body systems.

Let us now consider the neutron-rich Li isotopes in Fig. 5.1. To the right of the stable isotopes, we find four nuclei. As we add more neutrons, we find that the half-life drastically reduces, to the point that in ^{10}Li the nuclear system is unbound. However, ^{11}Li does bind again, and has a comparatively long half-life of 8.6 ms. This is an intriguing situation; why does the addition of *two* neutrons to the ^{9}Li system bind (briefly) into ^{11}Li, but not the addition of only a single neutron? ^{11}Li is now considered to be part of a class of nuclei, of which there are a number of such examples, that are known as 'halo' nuclei. These systems typically have very large radii compared to their neighboring isotopes. In the case of ^{11}Li, the dimension of the system is about the same size as that of ^{208}Pb [3]. In Fig. 5.2 we display another halo nucleus, ^{6}He (furthermore, compare the half-lives of ^{5}He and ^{7}He in Fig. 5.1). But what is the underlying physics for such systems? We now believe that these systems exhibit such behavior due to the NNN force, by seeing that the two halo neutrons interact very weakly with a third nucleon in the ^{9}Li 'core'. Note that two neutrons on their own, do not form a stable system, but the addition of the third nucleon, stabilizes the system (a signature of the three nucleon force).

The ^{11}Li system raises another interesting question: How many neutrons could one add to the Lithium system and still have some resemblance of a bound system? Such studies attempt to determine where the neutron drip line is located. The neutron drip line is the location on the nuclear chart at which the addition of any more neutrons to the system can no longer lead to additional binding energy. Thus, we loosely say that the neutrons drip out of the nucleus at that stage. Experimentally, the location of the neutron drip line, is, in general, not known precisely.

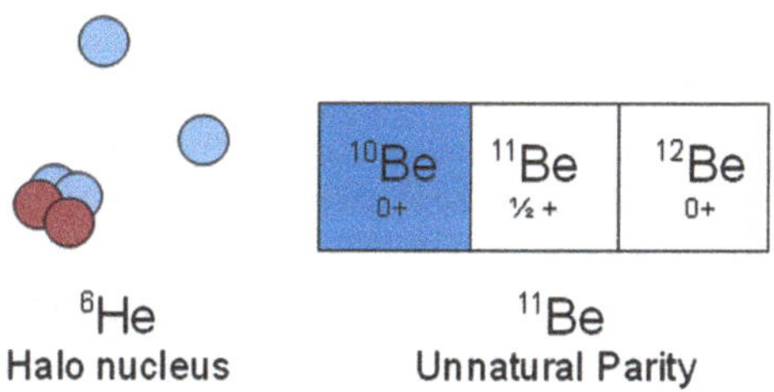

Fig. 5.2 To the *left*, we show a schematic picture of the ^{6}He halo nucleus. Note how the two neutrons 'orbit' the ^{4}He core at a large distance. To the *right* we show the gs parity assignment of three Be nuclei. ^{11}Be is expected to have a negative-parity gs, but instead has a positive-parity assignment. This is called an 'unnatural' parity assignment

Although the determination of the neutron drip line is important, many 'interesting' observations are already made in neutron-rich systems. One particular observation that we want to highlight is that of unnatural gs parity assignments. Consider Fig. 5.2, in which we show three Be isotopes, $^{10-12}$Be. Since Be is a p-shell nucleus, we expect that gs parity assignment will be positive for even-A isotopes and negative for odd-A isotopes. However, in the case of ^{11}Be, the gs parity assignment is positive (instead of negative). Theoretically, this is understood as the shell structure changing as the neutron to proton ratio increases. This particular observation of 'unnatural' gs parity assignments will be a major theme of this chapter, thus, we will leave the detailed discussions for Sect. 5.2.

5.1.2 The NCSM and Exotic Nuclei

We now turn our attention to the theoretical challenges in calculating the properties of neutron-rich nuclei, or generally exotic nuclei. From an overarching perspective, nuclei exhibit bound states, resonance states and also scattering states. Furthermore, they can undergo dynamic changes such as $\beta-$decay, or break apart into smaller fragments (fission). If we truly want to develop an *ab-initio* theory of nuclear physics, we need to be able to reliably describe all of these properties as well as supply an estimate to the uncertainty in our methods. Needless to say, this is a tough challenge, but nevertheless, worthy of further investigation.

The NCSM is very well suited for calculating the bound state properties of nuclei. This is especially true for states such as the ground state, as well as the low lying excited states. However, the theoretical uncertainty increases rapidly once we try to describe loosely-bound nuclei, such as those found away from the valley of stability. Why is this so?

The NCSM wavefunctions are expanded as a sum of anti-symmetric HO basis states. However, the asymptotic form of the HO basis falls off as a Gaussian $(\exp(-\frac{r^2}{b^2}))$, whereas the true asymptotic form of a bound-state actually has an exponential tail $\exp(-cr)$. In order to correct for this deficiency in the NCSM, one needs to extend the NCSM calculations to very large $N_{\max}$ spaces. Practically, we find 'a very large' $N_{\max}$ space to be computationally impossible. Thus, by using a limited $N_{\max}$ space, we are forced to make extrapolations to $N_{\max} = \infty$, leading to large theoretical uncertainties.

Yet, even if we were able to obtain suitably large $N_{\max}$ spaces from the outset, we still would not be able to describe resonance or scattering states properly (or realistically). That is because, these states lie in the continuum and, thus, must be treated so as to account for this fact. Furthermore, describing observables, such as phase-shifts, in order to determine where a resonance is located, is impossible, since we have no easy control of the dynamics. A new approach is needed.

5.1.3 Do We Really Need an **Ab-Initio** *Theory of Reactions?*

There has been a concentrated effort in the experimental community to build 'bigger and better' experimental facilities with the purpose of measuring the properties of neutron-rich nuclei (FRIB [4] and ARIEL [5] are two examples). Naturally, the ability to study reactions is also inherent in the experimental facilities.

In the previous section, we presented some of the issues, with which one is faced, when attempting a theoretical description of exotic nuclei or even more generally, dynamic changes in nuclei such as those encountered in reactions. Given the great advancements that have been made in experimental facilities, does one *really* need an *ab-initio* description of nuclear reactions? After all, why not simply measure the reaction pathways and be done with it. Admittedly, this is a very pessimistic view of theoretical nuclear reactions. However, it turns out that some of these experiments are plagued with all sorts of difficulties; we will highlight one of them to illustrate that theoretical calculations, besides improving our understanding of reactions, can actually complement the picture that experimental data have supplied.

A recent success of an *ab-initio* description of a nuclear reaction, *i.e.*, $d + {}^3\mathrm{He} \rightarrow p + {}^4\mathrm{He}$, illustrates the importance of theory [6]. One of the quantities of interest is the astrophysical $S-$factor, which is contained in the definition of the cross section, $\sigma(E) \sim S(E)E^{-1}$. The $S-$factor contains all the essential nuclear physics of the reaction, whereas the remaining terms in the cross section are dependencies on the energy.

Usually, one is interested in the astrophysical $S-$factor at very low energies. Determining the $S-$factor experimentally at low energies (on the order of tens of keV) is very difficult to do, since electron-screening effects occurring between the beam and the target artificially increase the size of the $S-$factor at those energies. Theoretical calculations on the other hand are free from electron-screening effects, and, thus, give the proper form of $S(E)$ at low energies. The difference between theory and experiment is quite noticeable at low energies (see Fig. 2 of [6]). In this way, experimental and theoretical efforts work side-by-side to build a better understanding of nuclear reactions.

5.1.4 The NCSM/RGM

The NCSM is well-suited to describing the bound-state properties of a nuclear system. Since we are using realistic interactions and performing the calculations from first-principles, no uncontrolled approximations are present. We would like to somehow extend the capabilities of the NCSM to the point where we can also calculate the properties of loosely bound nuclei, as well as being able to describe dynamic processes, such as reactions. The required physics is contained in the continuum, to which the NCSM has no access.

We will now present a way to merge the two techniques, the NCSM and the resonating group method (RGM); the result of combining the two bases is the NCSM/RGM [2, 7]. The RGM is particularly well-suited for describing the interaction amongst clusters of nuclei, in a setting such as scattering processes [8]. Naturally, we aim to combine the two techniques in a suitable manner for our purposes. Thus, to recap, the realistic nuclear interactions as well as the bound-state properties enter through the NCSM, whereas the continuum physics is contained in the RGM.

We will now briefly present the key ideas of the RGM aspect. We will present the binary cluster formulation of the RGM, although there is significant effort underway to extend the NCSM/RGM to ternary clusters. Before we present the mathematical formulation, we would like to present a physical picture. In the binary cluster formulation, one considers a heavy target nucleus, such as ^{8}He, as well as a lighter projectile nucleon (or nucleus), such as the neutron. To access the gs properties of the ^{9}He system, we calculate within the framework of the NCSM/RGM the scattering of the projectile off the target nucleus. In other words, we access the gs properties of ^{9}He through the scattering process $n + {}^8$He.

We begin with a basis consisting of binary clusters of total angular momentum J, parity π and isospin T.

$$|\Phi^{J^\pi T}_{\nu r}\rangle = \Big[\big(\big|A{-}a\,\alpha_1 I_1^{\pi_1} T_1\big\rangle \big|a\,\alpha_2 I_2^{\pi_2} T_2\big\rangle\big)^{(sT)} \times Y_\ell\left(\hat{r}_{A-a,a}\right)\Big]^{(J^\pi T)} \frac{\delta(r - r_{A-a,a})}{r r_{A-a,a}}. \tag{5.1}$$

In the above expression, $\big|A{-}a\,\alpha_1 I_1^{\pi_1} T_1\big\rangle$ and $\big|a\,\alpha_2 I_2^{\pi_2} T_2\big\rangle$ correspond to the two anti-symmetric cluster states. The former cluster represents the (heavier) target nucleus, composed of $A - a$ nucleons, whereas the (lighter) projectile consists of a nucleons. In the case of a single-nucleon projectile, $a = 1$. Each cluster state is specified by an intrinsic angular momentum I_i, parity π_i and isospin T_i, as well as additional quantum numbers α_i, where i indicates the cluster. The clusters are coupled to intermediate angular momentum $s = I_1 + I_2$ and total isospin $T = T_1 + T_2$. The spherical harmonic $Y_\ell\left(\hat{r}_{A-a,a}\right)$ represents the relative orbital angular momentum of the center of mass motion of the cluster.

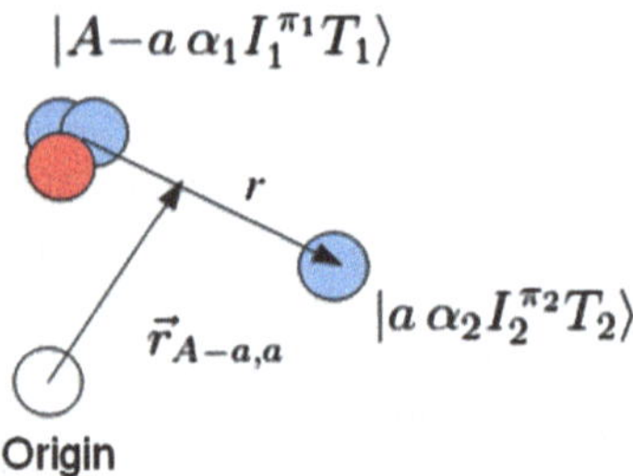

Fig. 5.3 A physical picture of Eq. (5.4) in which we indicate the separation of the two clusters by r and the relative motion of the center of mass is given by $\vec{r}_{A-a,a}$. Note that the coordinates are defined from a point in space, which we labeled as the 'origin' of the coordinate system. The cluster eigenstates are $|A-a\,\alpha_1 I_1^{\pi_1} T_1\rangle$ and $|a\,\alpha_2 I_2^{\pi_2} T_2\rangle$

$$\vec{r}_{A-a,a} = r_{A-a,a}\hat{r}_{A-a,a} = \frac{1}{A-a}\sum_{i=1}^{A-a}\vec{r}_i - \frac{1}{a}\sum_{j=A-a+1}^{A}\vec{r}_j\,, \tag{5.2}$$

where $\{\vec{r}_i, i = 1, 2, \cdots, A\}$ are the A single-particle coordinates. To simplify the notation, we group the cumulative quantum numbers into a single index $\nu = \{A-a\,\alpha_1 I_1^{\pi_1} T_1;\; a\,\alpha_2 I_2^{\pi_2} T_2;\; s\ell\}$.

The many-body wavefunction can now be expanded in terms of the binary-cluster basis states,

$$|\Psi^{J^\pi T}\rangle = \sum_\nu \int dr\, r^2 \frac{g_\nu^{J^\pi T}(r)}{r}\, \hat{\mathcal{A}}_\nu\, |\Phi_{\nu r}^{J^\pi T}\rangle\,. \tag{5.3}$$

Since we are dealing with nucleons, we require that the many-body wavefunction has to be antisymmetric under an exchange of *any* two nucleons. The cluster states themselves are antisymmetric, but the combination of the two as written in the basis $|\Phi_{\nu r}^{J^\pi T}\rangle$ is not anti-symmetric when two nucleons belonging to different clusters are exchanged. The anti-symmetry requirement is taken care of by introducing an intercluster anti-symmetrizer $\hat{\mathcal{A}}_\nu = \sqrt{\frac{(A-a)!a!}{A!}}\sum_P (-)^p P$, where the sum runs over all possible permutations P that can be carried out among nucleons pertaining to different clusters, and p is the number of interchanges characterizing them. The expansion coefficients in Eq. (5.3) are the relative-motion wavefunctions $g_\nu^{J^\pi T}(r)$, which represent the only unknowns of the problem. To determine them one has to solve the non-local integro-differential coupled-channel equations

$$\sum_\nu \int dr\, r^2 \left[\mathcal{H}_{\nu'\nu}^{J^\pi T}(r', r) - E\,\mathcal{N}_{\nu'\nu}^{J^\pi T}(r', r)\right] \frac{g_\nu^{J^\pi T}(r)}{r} = 0. \tag{5.4}$$

In order to have a physical picture in mind, we refer the reader to Fig. 5.3, in which we show some of the quantities given in Eq. (5.4).

The Hamiltonian kernel,

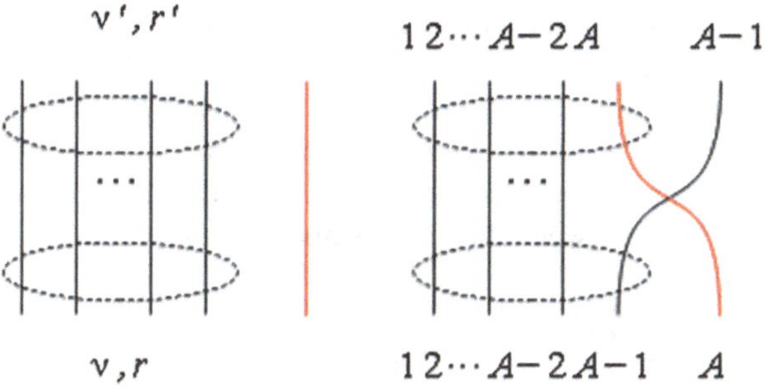

Fig. 5.4 The Norm kernel as shown in Eq. (5.6). The term on the *left* shows the direct term whereas the term on the *right* shows the exchange term. The *red line* indicates the light projectile whereas the *black lines* indicate nucleons inside the heavy target

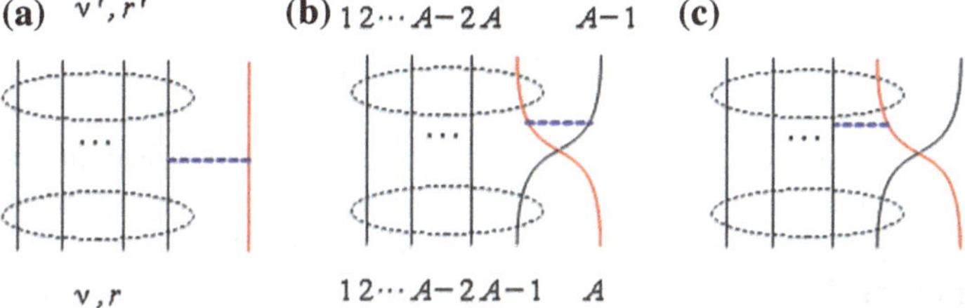

Fig. 5.5 The Hamiltonian kernel as shown in Eq. (5.5). Terms **a** and **b** represent the direct terms whereas term **c** represents the exchange term. The interaction between the clusters is given by the *dashed blue line*

$$\mathcal{H}^{J^\pi T}_{\nu'\nu}(r', r) = \left\langle \Phi^{J^\pi T}_{\nu' r'} \right| \hat{\mathcal{A}}_{\nu'} H \hat{\mathcal{A}}_{\nu} \left| \Phi^{J^\pi T}_{\nu r} \right\rangle , \tag{5.5}$$

and the norm kernel,

$$\mathcal{N}^{J^\pi T}_{\nu'\nu}(r', r) = \left\langle \Phi^{J^\pi T}_{\nu' r'} \right| \hat{\mathcal{A}}_{\nu'} \hat{\mathcal{A}}_{\nu} \left| \Phi^{J^\pi T}_{\nu r} \right\rangle , \tag{5.6}$$

contain the antisymmetrization as well as the nuclear-structure of the problem. We have used the notation E and H to denote the total energy in the center of mass frame, and the intrinsic A nucleon Hamiltonian, respectively. The kernels themselves have a physical interpretation, in terms of direct and exchange terms. In particular, the kernels can be expressed diagrammatically as shown in Fig. 5.4 for the norm kernel and Fig. 5.5 for the Hamiltonian kernel. The diagrams represent the required matrix elements arising from evaluating the antisymmetrizer; these terms must be derived individually. The evaluation of the kernels themselves is extensively discussed in [7].

The Hamiltonian contained in Eqs. (5.4) and (5.5) can be written as

$$H = T_{\rm rel}(r) + \mathcal{V}_{\rm rel} + \bar{V}_{\rm C}(r) + H_{(A-a)} + H_{(a)} , \tag{5.7}$$

where the last two terms are the intrinsic Hamiltonians for the target and projectile cluster. The cluster Hamiltonians are diagonalized in the NCSM, using the same $N_{\max}$, $\hbar\Omega$ space, and give us the cluster states $\left|A{-}a\,\alpha_1 I_1^{\pi_1} T_1\right\rangle$ and $\left|a\,\alpha_2 I_2^{\pi_2} T_2\right\rangle$. Note that softened realistic interactions are used, which require no further renormalization. $T_{\text{rel}}(r)$ and $\mathcal{V}_{\text{rel}}$ describe the relative kinetic energy of the clusters and the sum of the intercluster interactions between the various nucleons. $\mathcal{V}_{\text{rel}}$ is explicitly given by,

$$\begin{aligned}\mathcal{V}_{\text{rel}} &= \sum_{i=1}^{A-a}\sum_{j=A-a+1}^{A} V_{ij} - \bar{V}_{\text{C}}(r)\\ &= \sum_{i=1}^{A-a}\sum_{j=A-a+1}^{A}\Big[V_N(\vec{r}_i-\vec{r}_j,\sigma_i,\sigma_j,\tau_i,\tau_j)\\ &\quad + \frac{e^2(1+\tau_i^z)(1+\tau_j^z)}{4|\vec{r}_i-\vec{r}_j|} - \frac{1}{(A-a)a}\bar{V}_{\text{C}}(r)\Big], \qquad (5.8)\end{aligned}$$

in which V_N represents the nuclear interactions. Note that $\mathcal{V}_{\text{rel}}$ contains both the point-Coulomb interaction as well as an average Coulomb contribution ($\bar{V}_{\text{C}}(r)$). The term $\bar{V}_{\text{C}}(r) = Z_{1\nu}Z_{2\nu}e^2/r$, where $Z_{1\nu}$ and $Z_{2\nu}$ are the charge numbers of the clusters in channel ν. One subtracts the average Coulomb contribution in $\mathcal{V}_{\text{rel}}$ in order to have the term be local even in the presence of Coulomb interactions; the Coulomb contributions now fall off as r^{-2} instead of r^{-1}. Note that the subtraction is mathematically cancelled out in the total Hamiltonian (Eq. 5.7) by the addition of $\bar{V}_{\text{C}}$, and thus, this averaging procedure has no overall effect on the total Hamiltonian.

The δ functions that appear in the localized parts of the kernels are replaced by their representation in the HO model-space, using the same ($N_{\max}$, $\hbar\Omega$) values as we do for the cluster eigenstates. This replacement is only done for the localized parts, whereas for the diagonal parts of the identity operator in the antisymmetrizers, the kinetic term and average Coulomb term are treated exactly.

The norm and Hamiltonian kernels are translationally invariant quantities and can, thus, be naturally derived in the NCSM Jacobi-coordinate basis. For larger nuclei, such as those in the p-shell, the Jacobi coordinate approach becomes unfeasible, since the antisymmetrization of the target-cluster many-body states is unmanageable. Thus, we resort to using single-particle coordinates (i.e., Slater Determinants) for the target cluster. Our basis in the RGM is now composed of the heavier cluster state being expressed in terms of Slater determinants, whereas the light projectile is expressed in Jacobi-coordinates,

$$\begin{aligned}|\Phi_{\nu n}^{J^\pi T}\rangle_{\text{SD}} = \Big[\Big(|A{-}a\,\alpha_1 I_1 T_1\rangle_{\text{SD}}\,|a\,\alpha_2 I_2 T_2\rangle\Big)^{(sT)}\\ \times\, Y_\ell(\hat{R}^{(a)}_{\text{c.m.}})\Big]^{(J^\pi T)} R_{n\ell}(R^{(a)}_{\text{c.m.}})\,. \qquad (5.9)\end{aligned}$$

In order to recover the translationally invariant matrix elements in the Slater determinant basis, we make use of the following expression,

$$
{}_{\mathrm{SD}}\Big\langle \Phi^{J^\pi T}_{\nu' n'} \Big| \hat{\mathcal{O}}_{\text{t.i.}} \Big| \Phi^{J^\pi T}_{\nu n} \Big\rangle_{\mathrm{SD}} = \sum_{n'_r \ell'_r, n_r \ell_r, J_r} \Big\langle \Phi^{J_r^{\pi_r} T}_{\nu'_r n'_r} \Big| \hat{\mathcal{O}}_{\text{t.i.}} \Big| \Phi^{J_r^{\pi_r} T}_{\nu_r n_r} \Big\rangle
$$
$$
\times \sum_{NL} \hat{\ell}\hat{\ell}' \hat{J}_r^2 (-1)^{(s+\ell-s'-\ell')} \begin{Bmatrix} s & \ell_r & J_r \\ L & J & \ell \end{Bmatrix} \begin{Bmatrix} s' & \ell'_r & J_r \\ L & J & \ell' \end{Bmatrix}
$$
$$
\times \langle n_r \ell_r N L \ell | 00 n \ell \ell \rangle_{\frac{a}{A-a}} \langle n'_r \ell'_r N L \ell | 00 n' \ell' \ell' \rangle_{\frac{a}{A-a}} . \tag{5.10}
$$

Here $\hat{\mathcal{O}}_{\text{t.i.}}$ represents any scalar and parity-conserving translational-invariant operator ($\hat{\mathcal{O}}_{\text{t.i.}} = \hat{\mathcal{A}}$, $\hat{\mathcal{A}} H \hat{\mathcal{A}}$, etc.) and $\langle n_r \ell_r N L \ell | 00 n \ell \ell \rangle_{\frac{a}{A-a}}$ are generalized HO brackets for two particles with the mass ratio $a/(A-a)$. The use of the SD basis is computationally advantageous and allows us to explore reactions involving p-shell nuclei, as done in the present work. Finally, in order to evaluate the integration kernels, we need both the one- and two-body densities of the target eigenstates. These expressions are shown in Eqs. (51–52) in [7].

5.1.5 Orthogonality and General Observables

The presence of the norm kernel in Eq. (5.4) represents the fact that the many-body basis is expanded in terms of non-orthogonal basis functions. Thus, Eq. (5.4) does not represent a system of multi-channel Schroedinger equations and the relative motion functions $g_\nu^{J^\pi T}(r)$ are not Schroedinger wavefunctions. The non-orthogonality is short-ranged and originates from the non-identical permutations of the intercluster antisymmetrizers (these are present in the norm-kernel). Asymptotically, the norm-kernel retains orthogonality,

$$
\mathcal{N}^{J^\pi T}_{\nu' \nu}(r', r) \rightarrow \delta_{\nu' \nu} \frac{\delta(r' - r)}{r' r} . \tag{5.11}
$$

At large distances the relative wavefunctions $g_\nu^{J^\pi T}(r)$ do satisfy the *same* asymptotic boundary conditions as the relative wavefunctions of a multi-channel collision theory. Thus, one can correctly calculate physically meaningful quantities, such as the energy eigenvalues. However, the internal structure of the relative wavefunctions are still affected by the short-range non-orthogonality; this can lead to problems, when one wants to calculate general observables, such as transition matrix elements.

In order to remedy the situation, one can introduce an orthogonal version of Eq. (5.4),

$$
\sum_\nu \int dr r^2 \Big[\mathbb{H}^{J^\pi T}_{\nu' \nu}(r', r) - E \delta_{\nu' \nu} \frac{\delta(r' - r)}{r' r} \Big] \frac{\chi_\nu^{J^\pi T}(r)}{r} = 0 , \tag{5.12}
$$

where $\mathbb{H}^{J^\pi T}_{\nu'\nu}(r', r)$ is the Hermitian energy-independent non-local Hamiltonian defined by

$$\mathbb{H}^{J^\pi T}_{\nu'\nu}(r', r) = \sum_{\gamma'} \int dy' y'^2 \sum_{\gamma} \int dy\, y^2 \mathcal{N}^{-\frac{1}{2}}_{\nu'\gamma'}(r', y')\, \bar{\mathcal{H}}^{J^\pi T}_{\gamma'\gamma}(y', y)\, \mathcal{N}^{-\frac{1}{2}}_{\gamma\nu}(y, r). \tag{5.13}$$

The Schroedinger wavefunctions $\chi^{J^\pi T}_{\nu}(r)$ are now the wavefunctions we have to solve for and are related to the old $g^{J^\pi T}_{\nu}(r)$ by the equation,

$$\frac{\chi^{J^\pi T}_{\nu}(r)}{r} = \sum_{\gamma} \int dy\, y^2 \mathcal{N}^{\frac{1}{2}}_{\nu\gamma}(r, y)\, \frac{g^{J^\pi T}_{\gamma}(y)}{y}. \tag{5.14}$$

The orthogonalized version of Eq. (5.4) is, thus, given by,

$$\left[\hat{T}_{\rm rel}(r) + \bar{V}_{\rm C}(r) - (E - E^{I_1^{\pi_1} T_1}_{\alpha_1} - E^{I_2^{\pi_2} T_2}_{\alpha_2})\right] \frac{\chi^{J^\pi T}_{\nu}(r)}{r}$$
$$+ \sum_{\nu'} \int dr'\, r'^2\, W^{J^\pi T}_{\nu\nu'}(r, r')\, \frac{\chi^{J^\pi T}_{\nu'}(r')}{r'} = 0, \tag{5.15}$$

where $E^{I_i^{\pi_i} T_i}_{\alpha_i}$ is the energy eigenvalue of the i-th cluster ($i = 1, 2$). The $W^{J^\pi T}_{\nu\nu'}(r, r')$ is the overall non-local potential between the two clusters and depends on the relative channel numbers (ν, ν') but not on the energy. These are the equations that are solved for both bound and scattering states, depending on the boundary conditions that are imposed.

5.2 A NCSM/RGM Study of ^{9}He

One of the interesting questions that we can immediately address within the NCSM/RGM formalism, is the occurrence of unnatural parity states in the $N = 7$ isotones, shown in Fig. 5.1. As a reminder, an isotonic chain such as ^{11}Be, ^{10}Li and ^{9}He, is one, in which the number of neutrons remain the same in each nucleus, while the number of protons varies.

The heavier Helium isotopes, $^{6-9}$He, are currently one of the few chain of isotopes accessible to both detailed theoretical and experimental studies. In the case of ^{9}He, the neutron to proton ratio is $N/Z = 3.5$, making it one of the most neutron extreme systems studied so far. ^{9}He is particularly interesting theoretically, since it is part of a series of $N = 7$ isotones, in which it is believed that intruder states from the 1s0d HO shell are pushed down in energy into the 0p shell, allowing for the possibility of a positive-parity ground state. In this regard, ^{11}Be is the most famous example of having an un-natural parity assignment for the ground state, which has

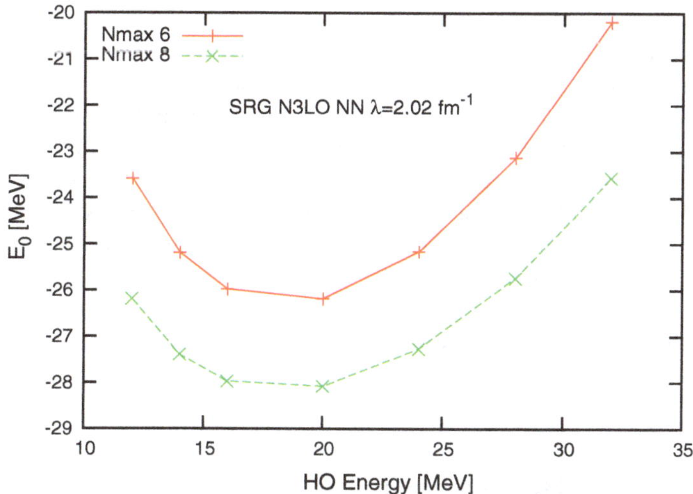

Fig. 5.6 The HO energy ($\hbar\Omega$) dependence of ^{8}He for the SRG N3LO NN interaction at $\lambda = 2.02\,\text{fm}^{-1}$. We expect for larger N_{max} values than shown, that the variational minimum will be located close to $\hbar\Omega \sim 16\,\text{MeV}$

been calculated theoretically [2, 9] as well as being observed experimentally [10]. Similarly, the same effect is seen in ^{10}Li [11].

Naturally, one might propose that the trend continues for ^{9}He. Shell model calculations have been attempted in the past [12], but no calculation of scattering lengths have been presented. Experimentally the situation is also debated. Early experiments found that the unbound ground state is a $\frac{1}{2}^-$ state [13–15]. These claims were challenged in [12], in which it was suggested that the ground state should be a $\frac{1}{2}^+$ virtual state; a claim further strengthened by their shell model calculations. Recently, the debate has been re-opened, in which two experiments were performed [16, 17], both claiming that the $\frac{1}{2}^+$ ground-state assignment is questionable. The authors of [17] suggest that the ground state should be reverted back to the previous $\frac{1}{2}^-$ assignment.

We present a study of the ^{9}He ground state in the framework of the IT-NCSM/RGM formalism [2, 7], by analyzing the $n-{}^8$He scattering process. The NCSM provides us with high-quality wavefunctions, obtained from a large-scale diagonalization of the Hamiltonian. We used the realistic chiral NN interaction [18], in which we have used all two-body terms up to next-to-next-to-next-to-leading order (N3LO). The interaction is then softened by using the SRG procedure with a momentum-decoupling scale of $\lambda = 2.02\,\text{fm}^{-1}$. The HO energy has been fixed at $\hbar\Omega = 16\,\text{MeV}$, as this is close to the variational minimum for ^{8}He. In our calculations, we can calculate the ^{8}He system up to $N_{\text{max}} = 10(11)$ in the full NCSM for positive (negative) parity states. However, we have also used the IT-NCSM to generate the ^{8}He wavefunctions for $N_{\text{max}} = 6(7) - 12(13)$ for positive (negative) parity states. This will serve as a good test of the IT-NCSM in an RGM setting.

As we have discussed, the NCSM has poor asymptotic behavior of the wavefunction for extended nuclei, due to the underlying HO basis. Thus, if one were to calculate ^{9}He directly in the NCSM, the calculation would lead to a rather poor description of the true ground state, especially if the system is loosely-bound. We will demonstrate that this is the case. The consequences of using the NCSM for loosely bound nuclei have already been demonstrated in [19], in which these problems are discussed. In order to improve the description of the asymptotic behavior, the NCSM calculation is coupled with the RGM. The RGM is particularly suitable for describing scattering processes, which includes the extra physics from the continuum that is lacking in the NCSM. As a final comment, note that we access the gs properties of ^{9}He by analyzing the phase shifts the neutron experiences, when we study the scattering of $n - {}^8$He.

5.2.1 IT-NCSM ^{8}He wavefunctions

In order to perform the NCSM/RGM calculations for $n - {}^8$He, we need to calculate the ^{8}He wavefunctions in a sufficiently large $N_{\max}$ basis space. The full NCSM calculations can be performed up to $N_{\max} = 10(11)$ for positive (negative) parity states, but are very time consuming. Instead, we made use of IT-NCSM calculations of the ^{8}He wavefunctions, in which we calculated the wavefunctions in the truncated $N_{\max}$ spaces up to $N_{\max} = 12(13)$. All IT-NCSM calculations have a complete $N_{\max} = 0 - 4$ basis space, and start the truncation of the basis at $N_{\max} = 6$. The harmonic oscillator energy of 16 MeV was chosen in the IT-NCSM calculations (See Fig. 5.6 for details). We outline the technical details of the IT-NCSM in the next two paragraphs.

The Positive-Parity States $J^\pi = \{0^+, 2^+\}$

We obtain the required wavefunctions from three sets of calculations. In order to obtain the required one- and two-body density matrix elements, we calculated the positive-parity states $J^\pi = \{0^+, 2^+\}$ for both $M = 0$ and $M = 1$. The basis for these two states is selected by the importance measure κ, in which we have used the extension to multiple reference states. In other words, if the importance measure for any one of the reference states (a previously calculated 0^+ or 2^+ state) satisfies the minimum threshold measure, then that basis state is kept in the overall basis for the positive-parity states. In the calculations for the positive-parity states, we used a minimum threshold of $\kappa_{\min} = 1.2 \times 10^{-5}$. The set of importance measures that we used to perform the extrapolations to $\kappa = 0$ for the $M = 0$ positive-parity states is given by $\kappa = \{5.5, 5.0, 4.5, 4.0, 3.5, 3.0, 2.5, 2.2, 2.0, 1.8, 1.4, 1.2\} \times 10^{-5}$. Note that there are 12 grid points. The calculated energies were extrapolated to $\kappa = 0$ by using either a cubic or quartic polynomial, and by using either the first- or second-

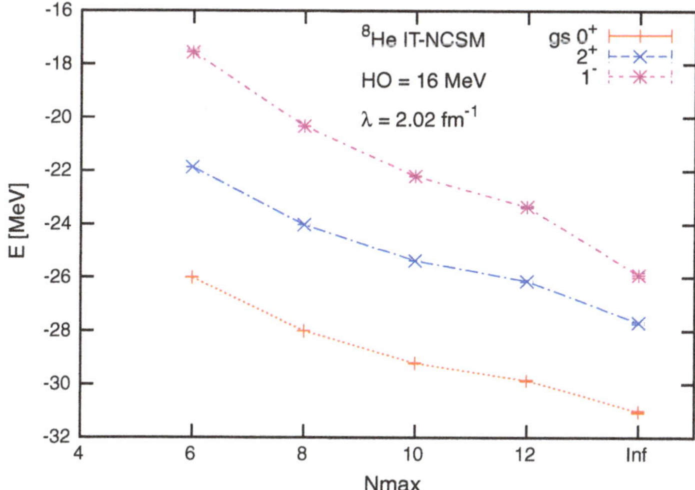

Fig. 5.7 The ^{8}He energy spectrum as calculated with the IT-NCSM procedure. We show *three states* that we are interested in; the 0^+ gs, and the excited states 2^+ and 1^-. We also show the exponential extrapolation to $N_{\text{max}} = \infty$ for each state, as well as indicating the uncertainty from the extrapolations in the IT-NCSM procedure

order results ($E^{(1)}_{0,\kappa}$ or $E^{(1+2)}_{0,\kappa}$). As a reminder, we refer to Sect. 3.2 and 3.4 for the details.

The result of using the first-order energies ($E^{(1)}_{0,\kappa}$), extrapolated to $\kappa = 0$ with a cubic polynomial, is shown in Fig. 5.7. We also show the extrapolation to $N_{\text{max}} = \infty$, in which we have used the standard exponential form, $f(N_{\text{max}}) = a * \exp(-b * N_{\text{max}}) + E_\infty$. Note that extrapolation takes into account the uncertainty from the extrapolations to $\kappa = 0$ for each N_{max} space.

In the case of the $M = 1$ positive-parity states, we only required the actual wavefunctions for the one- and two-body densities; thus, we only calculated the wavefunctions for the 0^+ and 2^+ state at $\kappa = \{3.0, 2.0, 1.2\} \times 10^{-5}$. We use a set of three different importance measures, as we would like to test the convergence in κ of NCSM/RGM calculations, when importance-truncated wavefunctions are used.

The Negative-Parity State $J^\pi = 1^-$

An older calculation, which was performed in the same manner as described above, also included the calculation of the $J = 1^+$ state. In the (IT-)NCSM calculations, the 1^+ state appears as the second excited state in the energy spectrum. However, experimentally, it is suggested that the second excited state in ^{8}He could be a negative-parity state, the $J = 1^-$ state (see Fig. 5.8).

The NCSM/RGM has the ability to take several excited states in the heavier target nucleus into account. These states are often physically relevant in obtaining an

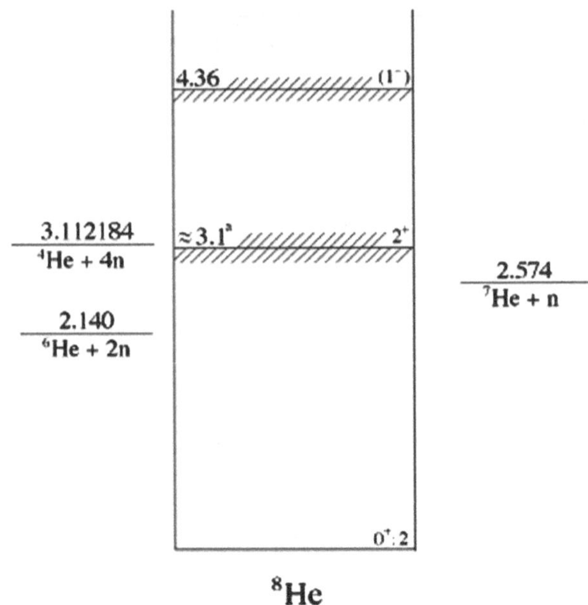

Fig. 5.8 The experimentally measured low-lying spectrum of the ^{8}He nucleus. States that are higher in energy are known, but not shown here. The figure has been adapted from [20]

accurate description of the system under investigation, in our case $n - {}^8$He. Thus, we have decided to also include the negative-parity state in the $n - {}^8$He calculation. As a reminder of the NCSM/RGM formalism, we refer the reader to Eq. (5.4), in which the inclusion of several target states, coupled to the angular momentum of the single nucleon, is denoted by the channel index $\nu = \{A{-}a\, \alpha_1 I_1^{\pi_1} T_1;\ a\, \alpha_2 I_2^{\pi_2} T_2;\ s\ell\}$.

The 1^- state is calculated in the IT-NCSM formalism, using only one reference state, in much the same way as the positive-parity states were calculated; the same set of κ-grid points were used. The $N_{\max}$ spaces are, however, not the same. In order to be consistent, we have to increase the $N_{\max}$ spaces for the negative-parity state by one unit from the corresponding positive-parity $N_{\max}$ space; otherwise, we are artificially limiting the $N_{\max}$ quanta for the negative-parity basis. For example, an $N_{\max} = 6$ positive-parity space will require an $N_{\max} = 7$ negative-parity space. This requirement can be understood from the alternating parity of the HO shells: Even-numbered major HO shells, such as $N = 0, 2, 4, \ldots$, are positive-parity shells, whereas the odd-numbered shells are all negative-parity shells.

Comparison of Various He Isotopes

Before we present our results on the NCSM/RGM calculation of $n - {}^8$He, we would like to illustrate the reliability of the SRG chiral NN N3LO interaction at

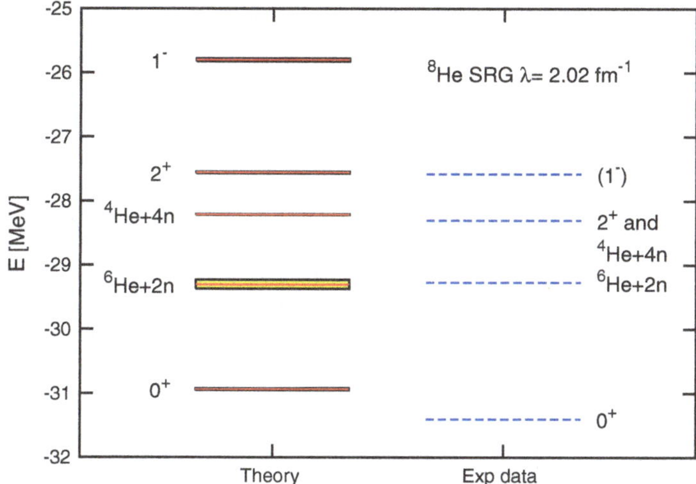

Fig. 5.9 The spectrum of IT-NCSM spectrum of ^{8}He, calculated using the SRG chiral N3LO interaction at $\lambda = 2.02\,\text{fm}^{-1}$ as compared to the experimentally known spectrum of ^{8}He. The energies shown for the IT-NCSM ^{8}He calculations are determined from the extrapolations to $N_{\text{max}} = \infty$ and are averaged over the various extrapolations that were performed to obtain the energies at $\kappa = 0$. We also show the ^{6}He and ^{4}He gs energies for comparison. $\hbar\Omega = 16\,\text{MeV}$ in all cases

$\lambda = 2.02\,\text{fm}^{-1}$, when it is used to calculate the energy spectrum of various Helium isotopes. In the experimental spectrum of ^{8}He, the neutron emission threshold of $n + {}^7$He is indicated, as well as the two-neutron emission of $2n + {}^6$He. The two-neutron emission threshold is the first decay channel (besides $\beta-$decay) that the ^{8}He nucleus can undergo.

Recently, the question was asked by S. Bacca whether or not our theoretically calculated ^{8}He nucleus is stable against two-neutron emission decay.[1] In order to be energetically stable against this emission process, we require that our gs energy for ^{8}He be lower than that of ^{6}He. We, thus, did a calculation of ^{6}He, using the same HO energy ($\hbar\Omega = 16\,\text{MeV}$) and the same interaction as before, and determined the gs energy. The result of these calculations are compared to the experimental spectrum of ^{8}He in Fig. 5.9.

The point of Fig. 5.9 is to show that we have calculated ^{8}He to be stable against two-neutron emission. Furthermore, we have shown that we agree reasonably well with the ^{8}He experimental spectrum when we use this particular interaction (chiral N3LO NN at $\lambda = 2.02\,\text{fm}^{-1}$), and use the HO energy of $\hbar\Omega = 16\,\text{MeV}$. The energies shown in the figure for the IT-NCSM ^{8}He calculations are determined from the extrapolations to $N_{\text{max}} = \infty$ and are averaged over the various extrapolations that were performed to obtain the energies at $\kappa = 0$. This leads to the 'bands' in the figure, indicating the uncertainty in our extrapolations. We also show the ^{6}He and

[1] Invited talk at TRIUMF January 2012.

^{4}He gs energies for comparison, as an indicator that our calculation is a reasonable assessment of the Helium isotopes.

5.2.2 The NCSM/RGM Calculation of $n - {}^{8}$He

The NCSM/RGM formalism requires the cluster state of the heavier target nucleus ($|A-a\,\alpha_1 I_1^{\pi_1} T_1\rangle$), in our case the ^{8}He wavefunctions. We also have to specify the energy of each state that is included in the NCSM/RGM calculation. In the results that we will present for the NCSM/RGM calculations, we will simply use the energy that corresponds to the wavefunction, at a specified value of κ. In other words, we do not use the extrapolated $\kappa = 0$ energies in the NCSM/RGM input. Typically, we will use the wavefunction and the associated eigenenergy corresponding to the smallest κ-threshold that we used, $\kappa_{\text{min}} = 1.2 \times 10^{-5}$.

Phase-Shift Calculations

In order to determine if there is a bound state in the ^{9}He system, we analyze the phase-shift of the $n - {}^{8}$He scattering calculation as a function of the kinetic energy of the neutron. An appearance of a resonance in the phase-shift at a specific energy usually indicates a physical state of some kind, for example, one that is found in the continuum. A structure-less phase-shift on the other hand would indicate that no physical states are present.

In Fig. 5.10 we show the NCSM/RGM calculated phase-shifts for the s- and p-wave channel. There is a resonance in the p-wave channel at about 1.7 MeV, which corresponds to a $\frac{1}{2}^{-}$ state. The s-wave channel shows no structure at all, except for a tiny 'bump' near the origin. This tiny feature might become relevant as the number of states included from the target nucleus increases; thus, we need to investigate the dependence of the phase-shift on the number of target states included. In Fig. 5.10a we show the phase-shifts for the target states of ^{8}He being the 0^{+} and 2^{+} states, whereas, in Fig. 5.10b, we show the phase-shifts, when the 1^{+} state is included, as well. It can be seen from the two figures that there is no significant difference in the phase-shifts if one includes the 1^{-} state; thus, since it is computationally demanding to include many of the possible target states, we will only include the positive-parity 0^{+} and 2^{+} states from now on.

In order to determine the nature of the s-wave phase-shift bump, we can determine the s-wave scattering length that corresponds to the calculated phase-shift. When the results, as shown in Fig. 5.10, are used, we determine that the s-wave scattering length[2] is $a_0(0^{+}, 2^{+}) = -1$ fm. The scattering length, in this case, is so small that we conclude that there is no bound state in the ^{9}He system. On the other hand, if the

[2] Nuclear physicists attribute a negative scattering length to an attractive potential, in contrast to the convention that particle physicists use.

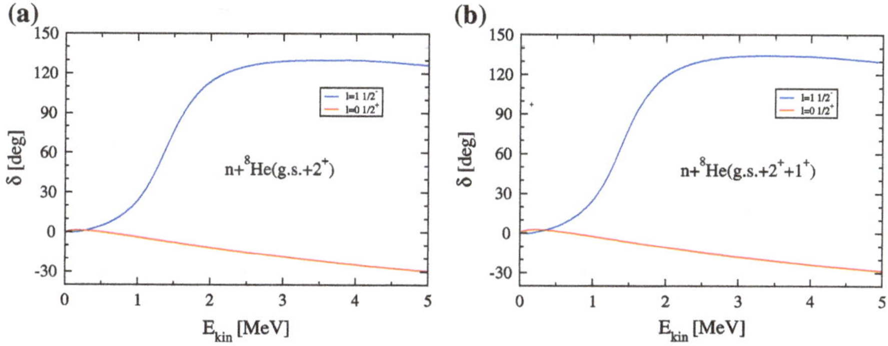

Fig. 5.10 The figures shows the calculated phase-shifts for the s- and p-wave channel at $N_{\max} = 12(13)$, obtained from IT-NCSM/RGM calculations. The phase-shift depends on the number of states included from the target nucleus. In **b** we show that the phase-shift is minimally affected by the inclusion of the 1^+ state, when compared to **a** in which only the 0^+ and 2^+ state were used

scattering length were significantly more negative, like on the order of ~ -10 fm, we could be dealing with a bound state.

We have also calculated the 1^- state in ^{8}He. This state is close in energy to the 1^+ state in the theoretical (IT-)NCSM calculations, and thus, could play a significant role in the NCSM/RGM phase-shift calculations. Furthermore, it is experimentally suggested that the second-excited state of ^{8}He is a 1^- state. In Fig. 5.11 we show the NCSM/RGM calculated s- and p-wave phase-shifts, in which we have included the target states $\{0^+, 2^+, 1^-\}$. Notice that the 'bump' in the s-wave phase-shift has increased significantly to a maximum of approximately 30 degrees. The phase-shifts are shown as a function of $N_{\max}$, in order to demonstrate that the NCSM/RGM calculations are converging in $N_{\max}$ and that the increase in the s-wave phaseshift near the origin is not an artifact of a specific $N_{\max}$ truncation. This structure in the s-wave phase shift signifies the presence of a physical state and, given that it would correspond to a $\frac{1}{2}^+$ gs, is very interesting for further study. On the other hand, the p-wave resonance remains largely unaltered with the inclusion of the 1^- state.

For the results shown in Fig. 5.11 we have determined the s-wave scattering length to be $a_0(0^+, 2^+, 1^-) = -12.59$ fm. This value is significantly more negative than our previously calculated scattering length (in which only positive-parity states were included in the target) of $a_0(0^+, 2^+) = -1$ fm. The large negative scattering length indicates that the true gs of the ^{9}He system is indeed a positive-parity $\frac{1}{2}+$ state.

Convergence of Phase-Shift Calculations

The difference seen in the s-wave phase-shifts between the positive-parity states and the inclusion of the 1^- state is quite striking. In order to gain confidence in the results that we have shown in Figs. 5.10 and 5.11, we need to demonstrate that the calculated phase-shifts are converged in some sense. We have already demon-

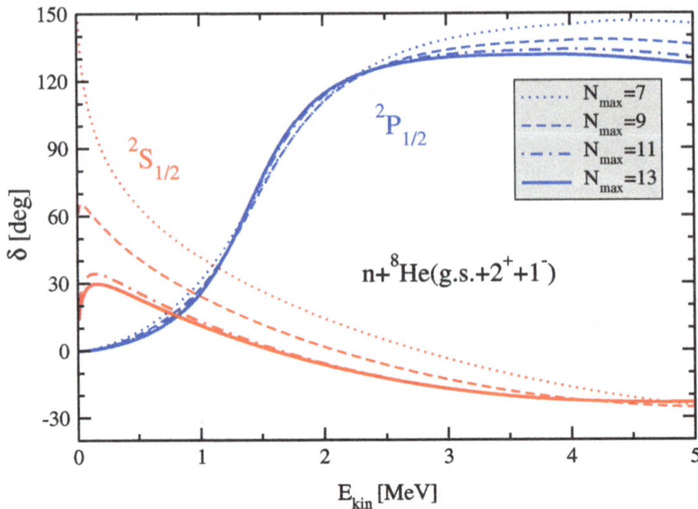

Fig. 5.11 The figure shows the (IT-)NCSM/RGM calculated s- and p-wave phase-shifts, in which the included target states are $\{0^+, 2^+, 1^-\}$, as a function of $N_{\max}$. The $N_{\max} = 12(13)$ calculations were obtained from an IT-NCSM calculation. Notice that the 'bump' in the s-wave phase-shift has increased significantly to a maximum of approximately 30 degrees in the largest $N_{\max}$ spaces. This structure in the s-wave phase shift signifies the presence of a physical state. The p-wave resonance remains largely unaltered with the inclusion of the 1^- state

strated the NCSM/RGM calculations have converged in $N_{\max}$ in Fig. 5.11. However, we still need to demonstrate that the NCSM/RGM calculations are reliable when the IT-NCSM ^{8}He wavefunctions are used. As an aside, this is a very interesting test of IT-NCSM calculated wavefunctions, compared to the usual tests, which are solely performed on the agreement between NCSM gs and IT-NCSM extrapolated gs energies. In other words, we are now truly testing the quality of the IT-NCSM wavefunctions.

We are specifically interested in the behavior of the s-wave phase-shift at low energies, since the structure in the phase-shift could be the signal of the true gs of the ^{9}He nucleus. Thus, all tests of convergence of the IT-NCSM wavefunctions will specifically focus on the converge of the s-wave phase-shift at low energies. In Fig. 5.12 we show the convergence in the importance measure κ of the s-wave phase-shift, in which we compare two different importance measures, $\kappa = 1.2 \times 10^{-5}$ and $\kappa = 2.0 \times 10^{-5}$, to the full NCSM calculations, which were performed at $N_{\max} = 10$. We can see from the figure that the s-wave phase-shift decreases as the importance measure is lowered, and, furthermore, as κ decreases, the calculated phase-shift tends towards the full NCSM result.

In Fig. 5.13, we show the κ-dependence on the s-wave phase-shift in the $N_{\max} = 12(13)$ space for three different importance measures, $\kappa = \{3.0, 2.0, 1.2\} \times 10^{-5}$. In this basis space, we are unable to perform full NCSM/RGM calculations, as it is too computationally demanding. We can see from the figure that the difference between

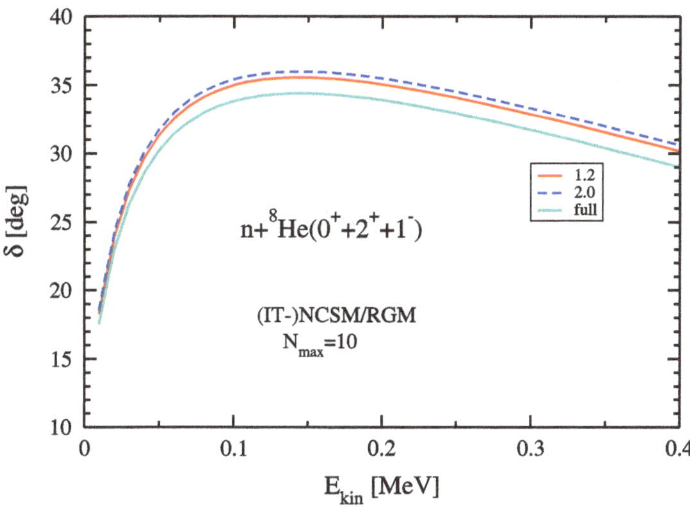

Fig. 5.12 The importance measure κ of the s-wave phase-shift, in which we compare two different importance measures, $\kappa = 2.20 \times 10^{-5}$ and $\kappa = 1.2 \times 10^{-5}$, to the full NCSM calculation. The phase-shifts shown correspond to the case when the $N_{\max} = 10(11)$ wavefunctions are used

the various s-wave phase-shifts decreases as the importance measure is decreased, indicating that we are converging in κ. Taking into account the difference between IT-NCSM and full NCSM results of the phase-shift, as shown in Fig. 5.12, as well as the convergence in κ, as shown in Fig. 5.13, we can conclude that the maximum value of the s-wave phase-shift at $N_{\max} = 12(13)$ is perhaps a little less than the 30 degrees. However, the 'bump' will still be present and, thus, does not alter our conclusions about the $\frac{1}{2}^{+}$ gs of ^{9}He.

Comparison to Experiment

In this subsection we return to the discussion on the experimental probing of ^{9}He. As we had mentioned earlier, there are essentially two classes of experimental results; a number of experiments claim that the gs should be a $\frac{1}{2}^{-}$ state, whereas the MSU experiment claims that the gs should be the $\frac{1}{2}^{-}$ state. The debate has not been settled yet, in part, due to the difficulty of the experiments themselves.

The experimental papers have [12, 16] also presented their calculated s-wave scattering lengths. These scattering lengths are determined from the experimental data. In the experiments of Al Falou, et al. the s-wave scattering length is estimated to be between $a_0 = -3 - 0$ fm. In the MSU experiment, the s-wave scattering length was deemed to be $a_0 < -10$ fm (i.e., more negative than -10 fm).

The results that have been presented in this section are, at the time of writing, only in a preliminary stage. Given the results that we have presented, we are in

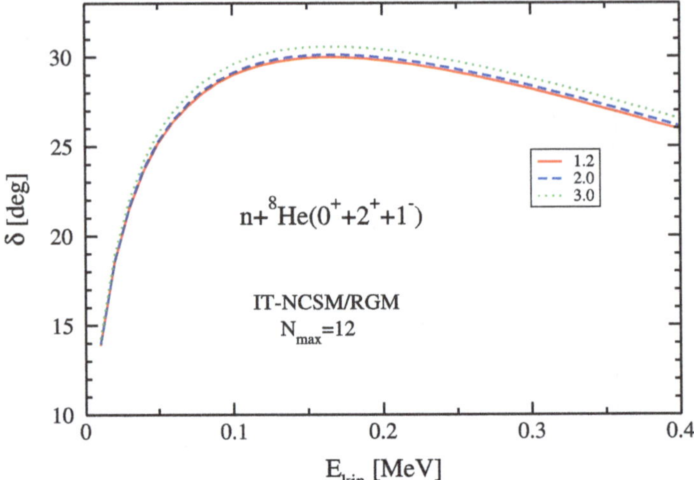

Fig. 5.13 The figures shows the κ-dependence on the s-wave phase-shift in the $N_{\max} = 12(13)$ space for three different importance measures, $\kappa = \{3.0, 2.0, 1.2\} \times 10^{-5}$. The difference between the various s-wave phase-shifts decreases as the importance measure is decreased, indicating that we are converging in κ

the unfortunate position of agreeing with neither one of the two main experimental results, namely, our calculated s-wave scattering length is neither $a_0 < -10$ fm nor $a_0 = -3 - 0$ fm, depending, on which sets of states are used from the ^{8}He target in the NCSM/RGM. However, it is always best to include as many of the states present in the target as possible. In this regard, we have more confidence in the calculations, which include the ^{8}He $0^+, 2^+, 1^-$ states (see Fig. 5.11) than the calculations that included only the positive-parity states from the target (see Fig. 5.10). Furthermore, we have shown that the phase-shifts are converging with both the size of the basis, $N_{\max}$, and with the decrease in the size of the importance measure, κ.

The experiments themselves are also very difficult to perform. To give the reader an idea of the nature of the experiments, we will briefly describe the MSU experiment that was performed in 2001 [21, 12]. Naturally, one might assume that all one needs to do is to have a beam of neutrons impinge on a ^{8}He target and then consider the outgoing nuclei that are formed. However, the half-life of the ^{8}He nucleus is 119 ms (see Fig. 5.1), which is too small to consider using ^{8}He targets directly; there is, of course, the difficulty of steering a neutron beam towards the target. Thus, the ^{9}He nucleus must be accessed by other means. This is an important consideration, which we now explain.

The MSU experiment considers the reaction ^{9}Be(^{11}Be,^{8}He + n)X, in which a beam of ^{11}Be nuclei at 28 MeV/u is incident on a ^{9}Be target. The direct-reaction leads to the production of ^{8}He + n and some other products (X). The neutron and the ^{8}He fragment are detected in coincidence, and are determined to be traveling at approximately the same velocity (within 1 cm/ns). Note that ^{9}He is not directly detected; it is the coincidence measurement of the neutron with the ^{8}He fragment that suggests that at some time prior to detection, the neutron and the fragment have interacted. The resulting distribution of the velocity difference between the fragment and the neutron is determined and is found to be approximately Gaussian in shape. The fact that the distribution is peaked and narrow suggests that the final-state interactions ($n-{}^8$He) are strong and are dominated by s-wave interactions. The experimental distribution of the velocity differences are then fitted to a theoretical potential model, based on some reasonable assumptions regarding the initial and final states of the reaction. The potential-model fit to the experimental data is quite involved, the details of which are not important here. However, it should be noted that the fit has a direct connection to the predicted s-wave phase-shift, a_0. By performing a number of fits to the experimental data, it is found that the experimental data favors a scattering length that is consistent with $a_0 < -10$ fm (*i.e.*, more negative than -10 fm). A scattering length of $a_0 \sim 0$ fm is clearly ruled out based on the value of the χ^2/dof of the obtained fits. It is important to note that the initial neutron state of ^{11}Be is dominated by a $1s_{1/2}$ single-particle orbital (recall the gs of ^{11}Be is, in fact, $\frac{1}{2}^+$). In other words, the initial neutron state originates in an s-state. This forms the basis of a selection rule argument in this particular experiment: since the neutron originates in an $l = 0$ state and is also detected with a velocity very near that of the fragment, it can be argued that the reaction product $n-{}^8$He is also in an $l = 0$ state.

Current Status of the Calculation

One intriguing possibility that remains is the presence of the 2^- state. In the NCSM calculations of ^{8}He, this state is located about 1 MeV higher in energy than the 1^- state, and, could play a role in the peak height of the low-energy s-wave phase-shift. Initial calculations in small $N_{\text{max}} \leq 6$ spaces, seem to suggest that the 2^- state lowers the height of the peak slightly. However, a thorough investigation at $N_{\text{max}} \geq 10$ is currently underway and will ultimately give a concrete answer as to the relevance of the 2^- state in the NCSM/RGM calculations.

5.2.3 *An IT-NCSM Calculation of* ^{9}He

To demonstrate that the NCSM leads to a poor description of loosely bound nuclei, we will now present some IT-NCSM calculations of ^{9}He. The IT-NCSM procedure is performed, as we have described numerous times before, so, we will only mention the key points. The ^{9}He $\frac{1}{2}^-$ and $\frac{1}{2}^+$ states are calculated using the chiral N3LO NN interaction, which has been softened by the SRG to $\lambda = 2.02\,\text{fm}^{-1}$. The HO energy is the same as for the ^{8}He case; $\hbar\Omega = 16$ MeV.

In Fig. 5.14 we show the result of the IT-NCSM calculations, in which we show the extrapolated $\kappa = 0$ energies for each N_{max} space. The extrapolations are the result of

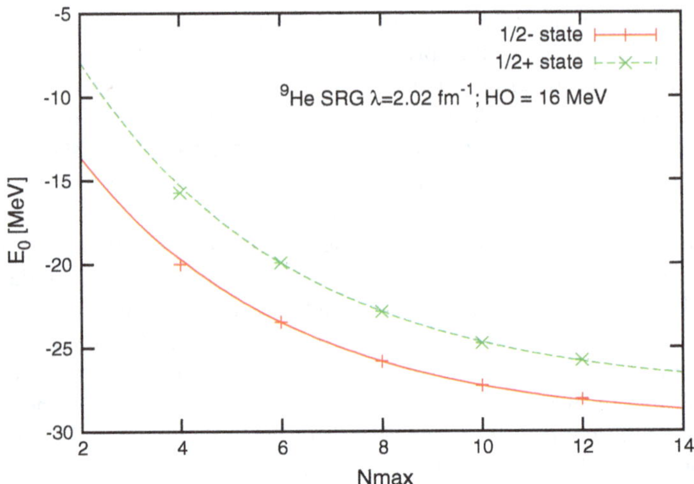

Fig. 5.14 The figure shows the IT-NCSM energies for the lowest $\frac{1}{2}^-$ and $\frac{1}{2}^+$ state of ^{9}He. Note that the gs is predicted to be the negative-parity state, instead of the positive-parity state, as predicted by the NCSM/RGM in $n-{}^8$He. The lines show the extrapolation to $N_{\text{max}}=\infty$; these are $E(\frac{1}{2}^-)=-29.568\pm0.044$ MeV and $E(\frac{1}{2}^+)=-27.698\pm0.063$ MeV

Table 5.1 A comparison of IT-NCSM and NCSM/RGM calculated energies of the $\frac{1}{2}^+$ and $\frac{1}{2}^-$ state in ^{9}He at $N_{\text{max}}=12(13)$

	IT-NCSM		NCSM/RGM	
state	$\frac{1}{2}^-$	$\frac{1}{2}^+$	$\frac{1}{2}^-$	$\frac{1}{2}^+$
E [MeV]	−28.040	−25.694	∼−28.317	∼−29.717

The approximate value of the NCSM/RGM is due to the uncertainty in the extrapolations of the ^{8}He gs at $N_{\text{max}}=12$

a cubic polynomial fitted to the first-order energies, $E^{(1)}_{0,\kappa}$. The uncertainty from the extrapolations are indicated, but are too small to be visible on the figure. Furthermore, we also show the extrapolation to $N_{\text{max}}=\infty$, using the standard exponential form, taking into account the uncertainties that arise from the IT-NCSM extrapolations. The exponential fit to determine the energies at $N_{\text{max}}=\infty$ only takes into account the energies at $N_{\text{max}}=6-12$.

We find that in the (IT-)NCSM calculations that the gs is predicted to be the $\frac{1}{2}^-$ state. This is in contrast to the prediction of the NCSM/RGM $n-{}^8$He calculation, in which the gs was predicted to be the positive-parity state $\frac{1}{2}^+$. The extrapolation to $N_{\text{max}}=\infty$ finds that the energies corresponding to the negative and positive-parity state are $E(\frac{1}{2}^-)=-29.568\pm0.044$ MeV and $E(\frac{1}{2}^+)=-27.698\pm0.063$ MeV, respectively.

We would like to make one final comparison between the IT-NCSM and IT - NCSM/RGM calculations of ^{9}He. In order to do this, we need to consider the calculations at $N_{\max} = 12(13)$, since the extrapolations to $N_{\max} = \infty$ were not performed for the NCSM/RGM calculations. In Table 5.1 we compare the absolute energies of the positive- and negative-parity states, as calculated in the IT-NCSM and with the NCSM/RGM. The IT-NCSM results come from an extrapolation to $\kappa = 0$ in the $N_{\max} = 12(13)$ space, in which a cubic polynomial was fitted to the first-order energies, $E^{(1)}_{0,\kappa}$. The NCSM/RGM values were calculated by first determining the gs energy of ^{8}He in $N_{\max} = 12(13)$, as calculated by the same extrapolation procedure for the IT-NCSM (cubic polynomial fitted to $E^{(1)}_{0,\kappa}$). We find that the gs energy of ^{8}He is, according to the extrapolation, $E_0(^8\text{He}) = -29.817$ MeV. To this energy we add 100 keV and 1.5 MeV to obtain the energy of the $\frac{1}{2}^+$ and $\frac{1}{2}^-$ state, respectively. These energies correspond to the structure seen in the phase-shift calculation of Fig. 5.11. Note that the splitting between the two states is much smaller in the NCSM/RGM case and that the NCSM/RGM predicts the $\frac{1}{2}^+$ state to be the gs of ^{9}He. Both sets of calculations, however, indicate that no bound state is present in ^{9}He; the calculated absolute energies are higher than the gs energy of ^{8}He.

References

1. K.M. Nollett, S.C. Pieper, R.B. Wiringa, J. Carlson, G.M. Hale, Quantum monte carlo calculations of neutron-α scattering. Phys. Rev. Lett. **99**, 022502 (2007)
2. S. Quaglioni, P. Navrátil, Ab Initio many-body calculations of $n-^3$H, $n-^4$He, $p-^{3,4}$He, and $n-^{10}$Be scattering. Phys. Rev. Lett. **101**, 092501 (2008)
3. R. Sánchez, W. Nörtershäuser, G. Ewald, D. Albers, J. Behr, P. Bricault, B.A. Bushaw, A. Dax, J. Dilling, M. Dombsky, G.W.F. Drake, S. Götte, R. Kirchner, H.-J. Kluge, Th. Kühl, J. Lassen, C.D.P. Levy, M.R. Pearson, E.J. Prime, V. Ryjkov, A. Wojtaszek, Z.-C. Yan, C. Zimmermann, Nuclear charge radii of ^{9,11}Li: The influence of halo neutrons. Phys. Rev. Lett. **96**, 033002 (2006)
4. http://www.frib.msu.edu/
5. http://www.triumf.ca/ariel
6. P. Navrátil, S. Quaglioni, Ab Initio many-body calculations of the ^{3}H$(d,n)^4$He and ^{3}He$(d,p)^4$He fusion reactions. Phys. Rev. Lett. **108**, 042503 (2012)
7. Sofia Quaglioni, Petr Navrátil, Ab initio many-body calculations of nucleon-nucleus scattering. Phys. Rev. C **79**, 044606 (2009)
8. Y.C. Tang, M. LeMere, D.R. Thompsom, Resonating-group method for nuclear many-body problems. Phys. Rep. **47**(3), 167–223 (1978)
9. I. Talmi, I. Unna, Order of levels in the shell model and spin of be^{11}. Phys. Rev. Lett. **4**, 469–470 (1960)
10. T. Aumann, A. Navin, D.P. Balamuth, D. Bazin, B. Blank, B.A. Brown, J.E. Bush, J.A. Caggiano, B. Davids, T. Glasmacher, V. Guimarães, P.G. Hansen, R.W. Ibbotson, D. Karnes, J.J. Kolata, V. Maddalena, B. Pritychenko, H. Scheit, B.M. Sherrill, J.A. Tostevin, One-neutron knockout from individual single-particle states of ^{11}be. Phys. Rev. Lett. **84**, 35–38 (2000)
11. M. Thoennessen, S. Yokoyama, A. Azhari, T. Baumann, J.A. Brown, A. Galonsky, P.G. Hansen, J.H. Kelley, R.A. Kryger, E. Ramakrishnan, P. Thirolf, Population of ^{10}Li by fragmentation. Phys. Rev. C **59**, 111–117 (1999)
12. L. Chen, B. Blank, B.A. Brown, M Chartier, A Galonsky, P.G. Hansen, M. Thoennessen, Evidence for an l = 0 ground state in 9he. Phys. Lett. B **505**(1–4), 21–26 (2001)

13. H.G. Bohlen, B. Gebauer, D. Kolbert, W. von Oertzen, E. Stiliaris, M. Wilpert, T. Wilpert, Spectroscopy of 9He with the (13C,13O)-reaction on 9Be. Z. Phys. A: Hadrons Nucl. **330**, 227–228 (1988) doi:10.1007/BF01293402.
14. K.K., Seth, M. Artuso, D. Barlow, S. Iversen, M. Kaletka, H. Nann, B. Parker, R. Soundranayagam, Exotic nucleus helium-9 and its excited states. Phys. Rev. Lett. **58**, 1930–1933 (1987)
15. W. von Oertzen, H.G. Bohlen, B. Gebauer, M. von Lucke-Petsch, A.N. Ostrowski, Ch. Seyfert, Th. Stolla, M. Wilpert, Th. Wilpert, D.V. Alexandrov, A.A. Korsheninnikov, I. Mukha, A.A. Ogloblin, R. Kalpakchieva, Y.E. Penionzhkevich, S. Piskor, S.M. Grimes, T.N. Massey. Nuclear structure studies of very neutron-rich isotopes of 710he, 911li and 1214be via two-body reactions. Nucl. Phys. A **588**(1), c129–c134 (1995). (Proceedings of the Fifth International Symposium on Physics of Unstable Nuclei)
16. H. Al Falou, A. Leprince, N.A. Orr, Investigation of continuum states in the unbound neutron-rich n = 7 and 9 isotones via breakup and knockout. Mod. Phys. Lett. A **25**(21–23), 1824–1827 (2010)
17. H.T. Johansson, Yu. Aksyutina, T. Aumann, K. Boretzky, M.J.G. Borge, A. Chatillon, L.V. Chulkov, D. Cortina-Gil, U. Datta Pramanik, H. Emling, C. Forssn, H.O.U. Fynbo, H. Geissel, G. Ickert, B. Jonson, R. Kulessa, C. Langer, M. Lantz, T. LeBleis, K. Mahata, M. Meister, G. Mnzenberg, T. Nilsson, G. Nyman, R. Palit, S. Paschalis, W. Prokopowicz, R. Reifarth, A. Richter, K. Riisager, G. Schrieder, H. Simon, K. Smmerer, O. Tengblad, H. Weick, M.V. Zhukov, The unbound isotopes 9,10he. Nucl. Phys. A **842**(14), 15–32 (2010)
18. D.R. Entem, R. Machleidt, Accurate charge-dependent nucleon-nucleon potential at fourth order of chiral perturbation theory. Phys. Rev. C **68**, 041001 (2003)
19. C. Forssén, P. Navrátil, W.E. Ormand, E. Caurier, Large basis ab initio shell model investigation of ^{9}Be and ^{11}Be. Phys. Rev. C **71**, 044312 (2005)
20. D.R. Tilley, J.H. Kelley, J.L. Godwin, D.J. Millener, J.E. Purcell, C.G. Sheu, H.R. Weller, Energy levels of light nuclei. Nucl. Phys. A **745**(34), 155–362 (2004)
21. L. Chen, Ph.D. thesis, Observation of initial and final-state effects in the systems (^{6}He + N), (^{8}HeN), and (^{9}Li + N). Michigan State University. Ph.D.Thesis (Advisor: G. Hansen) 2001

Chapter 6
Conclusion

The work in this thesis has presented several possible extensions to the No-Core Shell Model (NCSM). To recap, the NCSM is a first-principles technique, particularly relevant for lighter nuclei with $A \leq 20$, which allows us to calculate the observable properties of nuclei. We consider this approach an *ab-initio* method, as the only input to the problem is the nuclear Hamiltonian, which nowadays has a reasonably good connection to the underlying theory of QCD in terms of the chiral interaction. Such a connection is quite important considering that we want to place theoretical low-energy nuclear physics on solid ground. Even though the current chiral interaction, as given by Entem and Machleidt (as well as others), has problems regarding the power counting scheme used, as well as being non-renormalizable, it is still a step in the right direction. Work is under way to correct the deficiencies of the current chiral interaction, by using a power-counting scheme that is consistent and by making the interaction renormalizable. Unfortunately, much work remains to be done; hence, these improved chiral interactions will probably not be availabe for use in NCSM calculations for several years.

Solving the nuclear Hamiltonian requires specifying a many-body basis; for heavier nuclei ($A > 4$), we prefer to use a basis of Slater determinants consisting of single-particle harmonic oscillator (HO) wavefunctions. Consequently, the NCSM depends on two parameters, $N_{\max}$ and $\hbar\Omega$, in which the former indirectly specifies the size of the many-body basis and the latter is a result of working in the HO basis. Note that the $\hbar\Omega$ dependence remains, even if the bare nuclear Hamiltonian is used. The combination of using the HO basis and truncating the basis according to energy-quanta levels $N_{\max}$ leads to the exact separation of center-of-mass states from the intrinsic states. However, the HO basis has a Gaussian tail, which, unfortunately, leads to slow convergence in the NCSM calculations. Furthermore, expressing the nuclear interactions in the HO basis also leads to slow convergence. Convergence in the NCSM is usually considered to mean that we have reached a 'large enough' basis ($N_{\max} \sim 10 - 16$), so that any additional increase in the $N_{\max}$ basis would numerically lead to the same calculated properties (energy etc.) of the ground-state (gs). For $A \leq 4$, convergence can be attained directly. In the case of incomplete

M. K. G. Kruse, *Extensions to the No-Core Shell Model*,
Springer Theses, DOI: 10.1007/978-3-319-01393-0_6,

© Springer International Publishing Switzerland 2013

convergence ($A > 4$), a series of calculated gs energies are extrapolated, as a function of $N_{\max}$, to obtain the gs energy at $N_{\max} = \infty$. In this regard, the Similarity renormalization group (SRG) interactions have improved the situation remarkably; the use of a phase-shift equivalent soft interaction retains the variational nature of bare NCSM calculations as well as reducing the required $N_{\max}$ in order to obtain the onset of convergence. However, the SRG procedure comes at the price of inducing higher-body terms (such as induced three-body forces) that must be considered if the procedure is to remain unitary. Furthermore, the size of induced four-body forces is currently under investigation [1].

In Chaps. 1 and 2 we described how the NCSM basis grows rapidly as $N_{\max}$ increases. This is particularly significant for nuclei that lie in the middle of the p-shell, where the combinatorial factors that arise from filling the single-particle states in the NCSM basis are maximal. Even when SRG interactions are used, the number of $N_{\max}$ spaces required is still to large to approach convergence satisfactorily.

6.1 IT-NCSM

The importance-truncated NCSM (IT-NCSM) uses an argument based on first-order, multiconfigurational theory to select a small number of basis states present in the large NCSM basis. The selection of a basis state ($|\phi_\nu\rangle$) that is to be included from the next-larger $N_{\max}$ space is determined by means of the importance measure, $\kappa = \frac{|\langle\phi_\nu|H|\Psi_{\text{ref}}\rangle|}{\epsilon_\nu - \epsilon_{\text{ref}}}$, in which $|\Psi_{\text{ref}}\rangle$ represents a physical state calculated in a previous (possibly truncated) $N_{\max}$ space. The two energy terms in the denominator represent the HO quanta of the basis state and the reference state, respectively. By controlling the size of κ, we are able to include more or fewer basis states in the truncated NCSM calculations. To recover the result of a full NCSM calculation, in which all the basis states are kept in a specific $N_{\max}$ space, we are required to extrapolate a series of gs energies, as a function of κ, to the case when $\kappa = 0$. Extrapolations inherently possess dependencies on the choice of the function or the number of data points used; in the case of the IT-NCSM, these uncertainties as well as the difference to full NCSM calculations have only been eluded to in the past. It was the purpose of Chap. 3 to provide a detailed study of IT-NCSM, in general.

In Chap. 3 we presented a detailed study of IT-NCSM calculations for ^{6}Li, in which we have studied the dependence of IT-NCSM on various parameters. These include the behavior of IT-NCSM as a function of the model-space $N_{\max}$, the HO energy $\hbar\Omega$, the extrapolating functions used for the two types of data sets ($E^{(1)}_{0,\kappa}$ or $E^{(1+2)}_{0,\kappa}$), the SRG momentum-decoupling scale λ, as well as the influence on the basis selection procedure, when multiple reference states are used. The IT-NCSM calculations were then compared to NCSM calculations, in the same $N_{\max}$ basis space, as a way to estimate the accuracy of the procedure. We find that the extrapolations used in IT-NCSM, using either the first- or second-order results ($E^{(1)}_{0,\kappa=0}$ or $E^{(1+2)}_{0,\kappa=0}$) give similar results, even when different extrapolating functions are used. At $N_{\max} = 14$ we find that the IT-NCSM extrapolated gs energies differ from the NCSM gs energies by about 100–150 keV, whereas in extrapolations to $N_{\max} = \infty$, the difference is about 250 keV. The IT-NCSM calculations show no quantitative difference for the two SRG momentum-decoupling scales we used ($\lambda = 2.02$ fm^{-1} and $\lambda = 1.50$ fm^{-1}), in terms of the extrapolation errors or in regard to the difference to the full NCSM calculations. Two new features were seen in these calculations that have not been reported before: (1) IT-NCSM calculations seem to deteriorate in quality as $\hbar\Omega$ increases, when the same κ-grid is used (see Figs. 3.8 and 3.12); (2) using several reference states leads to a better basis selection for the gs energy than using just a single reference state (see Fig. 3.16). We propose that future IT-NCSM calculations should use multiple reference states to select the basis states, use smaller κ threshold limits as $\hbar\Omega$ increases and provide a reasonable error estimate of the $N_{\max} = \infty$ extrapolated energies.

In terms of matter radius results, we caution the reader that the extrapolation techniques used to extract general observables, other than the gs energy, are far less developed. In fact, it is worth reiterating that the convergence, as $N_{\max}$ increases for general operators does not necessarily follow a variational pattern and must be treated on a case by case basis. In our matter radius extrapolations to $N_{\max} = \infty$, we were fortunate to see a regular pattern as $N_{\max}$ increases which allowed us to extrapolate to $N_{\max} = \infty$. It is very encouraging to see that we could reliably calculate the matter radius operator in the IT-NCSM to at least within the accuracy of experimental techniques. This is particularly good news, considering that the operator is of a long-range nature. It would be very interesting to see how electromagnetic operators, particularly how the quadropole moment and BE(2) operators behave in an IT-NCSM setting.

We share the opinion that the criticism of importance truncated calculations is worth addressing. In this regard, we presented an extensive discussion on size-extensivity issues in many-body physics, in general, as well as how the issue arises in iterative IT-NCSM calculations, in particular. Nowadays, we prefer to use the sequential IT-NCSM formalism, which in addressing the original criticism that IT-NCSM lack size-extensivity, does include all particle-hole excitations up to the required $N_{\max}$ level. On the one hand, every test, to which IT-NCSM has been subjected, has come within a few hundred keV of the exact NCSM result. On the other hand, no conclusive evidence exists that sequential IT-NCSM is indeed size-extensive. In order to truly answer this question, one would need to demonstrate that IT-NCSM recovers the result of the NCSM over a wide range of mass numbers. Given the difficulty of performing full NCSM calculations, we might only be able to address the issue of size-extensivity for heavier nuclei, calculated with IT-NCSM, in a limited capacity.

6.2 NCSM Extrapolations

For heavier nuclei ($A > 4$) we typically find that the gs energy has not yet converged in the $N_{\max} \sim 10$ space. However, with the use of soft interactions, such as those generated by the SRG procedure, it is possible to perform an extrapolation to $N_{\max} = \infty$. Several extrapolation schemes have been proposed in past calculations [2–4]. All extrapolations that are used assume a decaying exponential form, in which the argument of the exponential assumes a dependence on either the size of the basis $N_{\max}$ or the UV/IR regulators, as we did in Chap. 4. The use of an exponential form has never been questioned. This is not surprising, since mathematical theorems, as well as explicit calculations in the case of Hyperspherical harmonics, have shown that smooth interactions, such as the (softened) chiral N3LO, generally converge exponentially as the basis increases [5, 6]. Thus, the use of an exponential form to extrapolate gs energies is warranted mathematically as well as practically.

Since we are extrapolating our calculated gs energies in a prescribed manner (either via $N_{\max}$ or by using UV/IR regulators) we are sensitive to uncertainties that arise in the extrapolations. Determining these uncertainties in a reliable manner has recently garnered some attention in the nuclear physics community.[1] This process goes by the name *uncertainty quantification*. It is important to note that uncertainty quantification is highly desirable if we want to place stringent error bars on our theoretical calculations. In this regard, we are starting to move away from quoting single, (presumably) exact gs energies of nuclei, obtained by diagonalization in a fixed $N_{\max}$ space, but instead are quoting numbers that are free from the NCSM parameters ($N_{\max}$ and $\hbar\Omega$) *and* that come with some uncertainties due to the residual effects of the NCSM parameters.

In this thesis we were not too concerned with determining rigid uncertainties in extrapolation procedures (except for those used in IT-NCSM, as in Chap. 3). Instead, we concerned ourselves with presenting a new extrapolation method based on the principles of effective field theory (EFT). We reformulated the traditional NCSM model-space of ($N_{\max}, \hbar\Omega$) into an EFT-style model-space, characterized by a UV and IR scale, (Λ, λ_{IR}). In the EFT model-space, we treat the UV and IR regulators on an equal footing. This is, in contrast, to the case of the traditional NCSM extrapolations, in which $N_{\max}$ is generally taken to be 'more important' than $\hbar\Omega$. Furthermore, in the EFT model-space, one has complete control over the UV and IR physics in the many-body problem. To this extent, a deficiency in a many-body calculation, as characterized by the size of the relative error $|\frac{\Delta E}{E}|$, can be attributed to excluded UV or IR physics—simply by raising the UV or lowering the IR scale, we should decrease the size of the relative error.

Since the discussions on the form of the IR regulator are still under debate, we refer the interested reader to the Conclusions section of Chap. 4 for a complete discussion on the issues regarding extrapolating with EFT regulators.

[1] James Vary's presentation at the Workshop on 'Perspectives on the NCSM', Vancouver 2011.

6.3 NCSM/RGM and Exotic Nuclei

The extension of the NSCM to larger $N_{\max}$ spaces by using a combination of IT-NCSM and extrapolation procedures is not always sufficient to accurately describe the physics of certain nuclei. The NCSM (and, thus, also IT-NCSM) is well-suited to describe the bound-state properties of nuclei, in which the first breakup channel is fairly high in excitation energy. This is true for nuclei, such as ^{4}He, where the breakup channel is located some 20 MeV above the gs energy, but not true for loosely bound nuclei or resonance states (e.g., the $A = 5$ systems). Loosely bound nuclei are usually located near the physically interesting region of the neutron dripline; a region where even the notion of conventional shell structure might seize to exist. As we have described before, one of the problems that the NCSM encounters in describing these exotic nuclei is due to the incorrect asymptotic behavior of the HO basis; bound states in nature fall of exponentially, whereas the HO basis has a Gaussian tail. The asymptotic nature of the basis is not the only issue to consider—the NCSM also has no way to describe scattering processes.

On the other hand, the resonating group method (RGM) is well-adapted to describe scattering phenomena between two clusters of nuclei. All that is required is to have a good description of the bound state properties of the clusters themselves, which is provided to us by the use of realistic interactions and the NCSM, if each cluster has a high breakup channel. By combining the NCSM with the RGM, we are able to extend the applicability of the NCSM to the edges of the nuclear chart.

In Chap. 5 we considered the interesting nucleus ^{9}He. This nucleus is part of a series of $N = 7$ isotones, which also includes ^{11}Be and ^{10}Li. It is believed that these ground-state have an unnatural-parity assignment. For ^{11}Be and ^{10}Li the unnatural-parity assignment has been confirmed. In the case of the odd-A ^{9}He nucleus, the natural gs parity assignment should be a negative-parity state. It has been suggested in different experiments that the gs state could be either a positive- or a negative-parity state. The binary-cluster formulation of the NCSM/RGM is well-adapted to calculating the gs properties of ^{9}He by considering the scattering process n–^{8}He. In our calculations we determine the gs to be a $\frac{1}{2}^+$ state. In order to gain more insight into the nature of the gs, we also calculated the scattering length, a_0, of the $\frac{1}{2}^+$ state. The scattering length that we determine is dependent on the number of states we include in the heavier target nucleus (^{8}He). In the case of including only the positive-parity ^{8}He states $J = \{0^+, 2^+, 1^+\}$, we determine the scattering length to be $a_0 = -1$ fm, in agreement with the experiments of [7, 8]. However, when we include the negative-parity state $J^\pi = 1^-$, we determine the scattering length to be $a_0 = -12.59$ fm, in agreement with the MSU experiment [9]. We are fairly certain that the ^{8}He $J^\pi = 1^-$ state must be included in the NCSM/RGM calculations, in part due to the large effect the state has on the phase-shift as well as scattering length calculations. In the ^{8}He IT-NCSM calculations one finds that a 2^- state is located within 1 MeV of the 1^- state. The relevance of this 2^- state in the phase-shift and scattering length calculations is currently under investigation.

In Chap. 5 we also demonstrated that a direct IT-NCSM calculation of ^{9}He leads to a prediction that the gs is a $\frac{1}{2}^-$ state. In this calculation, the positive-parity state $\frac{1}{2}^+$ is located about 2.5 MeV above the $\frac{1}{2}^-$ state. Such behavior is due to the slow asymptotic convergence of the NCSM with respect to loosely bound or resonance states and demonstrates the need for the NCSM/RGM formalism.

6.4 A Final Word

It was my aim to convey to the reader a good understanding of extensions to the NCSM. These extensions allow for the possibility to study heavier as well as exotic nuclei in the p-shell and could be used in describing sd-shell nuclei. However, it is important to realize the limitations of these methods. The NCSM and any of its various extensions will never supply a complete description of the nuclear chart; traditional shell model and configuration-interaction approaches are more suitable for heavier nuclei. It can, however, be used to gain valuable insight into the nature of nuclear forces, especially in terms of the role of three-body forces. Recently, there has been some curiosity into whether or not the current nuclear interactions (chiral or Av18+UIX) adequately describe nuclear systems with a high degree of isospin asymmetry (i.e., neutron-rich systems) [10]. Calculations such as those require a great deal of effort and should be performed. The NCSM is certainly well-adapted to answer such questions.

I find it worthwhile to hint at future research projects before I conclude this thesis. A potential area of study is the chain of Oxygen isotopes, in which we plan to determine where the neutron-drip line is located. Such calculations have been attempted before [11, 12], but they typically lack the inclusion of the three-nucleon force directly.[2] Furthermore, since we are using the chiral EFT potential, which has a direct connection to QCD, we can, in principle, determine which features of the interaction are enhancing (or reducing) the stability of the nucleus, as excess neutrons are added. For instance, is extra stability gained from three-nucleon forces or not? If so, which terms in the interaction are responsible for the stability, or the lack thereof?

Perhaps most importantly, can we determine *theoretically*, where the drip-line is located for other nuclei, before experimental data becomes available? These data are expected to come from the *Facility for Rare Isotope Beams (FRIB)*, which is currently under construction at Michigan State University. Having the ability to theoretically predict properties of nuclei (within some quoted uncertainties) *before* the experiments are performed would be a significant step forward for the theoretical nuclear-structure community.

[2] Instead, density-dependent two-body forces are used to mimic the nature of NNN forces.

References

1. R. Roth, J. Langhammer, A. Calci, S. Binder, P. Navrátil, Similarity-transformed chiral $nn+3n$ interactions for the Ab Initio description of ^{12}**C** and ^{16}**O**. Phys. Rev. Lett. **107**, 072501 (2011). http://link.aps.org/doi/10.1103/PhysRevLett.107.072501
2. C. Forssén, P. Navrátil, W.E. Ormand, E. Caurier, Large basis ab initio shell model investigation of ^{9}Be and ^{11}Be. Phys. Rev. C **71**, 044312 (2005). http://link.aps.org/doi/10.1103/PhysRevC.71.044312
3. C. Forssén, J.P. Vary, E. Caurier, P. Navrátil, Converging sequences in the ab initio no-core shell model. Phys. Rev. C **77**, 024301 (2008). http://link.aps.org/doi/10.1103/PhysRevC.77.024301
4. P. Maris, J.P. Vary, A.M. Shirokov, Ab initio no-core full configuration calculations of light nuclei. Phys. Rev. C **79**, 014308 (2009). http://link.aps.org/doi/10.1103/PhysRevC.79.014308
5. I.M. Delves, Variational techniques in the nuclear three-body problem. Adv. Nucl. Phys. **5**, 1–224 (1972)
6. T.R. Schneider, Convergence of generalized spherical harmonic expansions in the three nucleon bound state. Phys. Lett. B **40**(4), 439–442 (1972). ISSN 0370-2693, http://www.sciencedirect.com/science/article/pii/037026937290545X
7. H. Al Falou, A. Leprince, N.A. Orr, Investigation of continuum states in the unbound neutron-rich n = 7 and 9 isotones via breakup and knockout. Mod. Phys. Lett. A **25**(21–23), 1824–1827 (2010)
8. H.T. Johansson, Yu. Aksyutina, T. Aumann, K. Boretzky, M.J.G. Borge, A. Chatillon, L.V. Chulkov, D. Cortina-Gil, U. Datta Pramanik, H. Emling, C. Forssn, H.O.U. Fynbo, H. Geissel, G. Ickert, B. Jonson, R. Kulessa, C. Langer, M. Lantz, T. LeBleis, K. Mahata, M. Meister, G. Mnzenberg, T. Nilsson, G. Nyman, R. Palit, S. Paschalis, W. Prokopowicz, R. Reifarth, A. Richter, K. Riisager, G. Schrieder, H. Simon, K. Smmerer, O. Tengblad,H. Weick, M.V. Zhukov, The unbound isotopes 9,10he. Nucl. Phys. A **842**, 15–32 (2010). ISSN 0375-9474, http://www.sciencedirect.com/science/article/pii/S0375947410004057
9. L. Chen, B. Blank, B.A. Brown, M. Chartier, A. Galonsky, P.G. Hansen, M. Thoennessen, Evidence for an l = 0 ground state in 9he. Phys. Lett. B **505**(1–4), 21–26 (2001). ISSN 0370-2693, http://www.sciencedirect.com/science/article/pii/S0370269301003136
10. E.A. McCutchan, C.J. Lister, Steven C. Pieper, R.B. Wiringa, D. Seweryniak, et al. On the lifetime of the 2 state in 10C. Phys. Rev. C **86**, 014312 (2012). arXiv:1201.2960[nucl-ex]
11. G. Hagen, M. Hjorth-Jensen, G.R. Jansen, R. Machleidt, T. Papenbrock, Continuum effects and three-nucleon forces in neutron-rich oxygen isotopes. Phys. Rev. Lett. **108**, 242501 (2012). arXiv:1202.2839[nucl-th]
12. Ø. Jensen, G. Hagen, M. Hjorth-Jensen, J.S. Vaagen, Closed-shell properties of ^{24}O with ab initio coupled-cluster theory. Phys. Rev. C **83**, 021305 (2011). http://link.aps.org/doi/10.1103/PhysRevC.83.021305

Curriculum Vitae

Michael Kruse
7000 East avenue
Livermore, CA 94550
USA
(✉) kruse9@llnl.gov

Education

2006–2012 **Ph.D. Physics**, *University of Arizona*, Tucson, USA.
2004–2006 **B.sc. Hons. Physics**, *University of Pretoria*, Pretoria, South Africa.
2001–2004 **B.sc. Physics**, *University of Pretoria*, Pretoria, South Africa.

Experience

July 2012 **Postdoc**, *Lawrence Livermore National Laboratory*, Livermore. present

Awards

- Galileo Scholarship award, April 2012
- Galileo Scholarship award, April 2009

Selected Publications

1. M.K.G. Kruse, E.D. Jurgenson, P. Navrátil, B.R. Barrett, W.E. Ormand, Extrapolation uncertainties in the importance-truncated no-core shell model. Phys. Rev. C **87**, 044301 (2013)

M. K. G. Kruse, *Extensions to the No-Core Shell Model*,
Springer Theses, DOI: 10.1007/978-3-319-01393-0,
© Springer International Publishing Switzerland 2013

2. S.A. Coon, M.I. Avetian, M.K.G. Kruse, U. van Kolck, P. Maris, J.P. Vary, Convergence properties of *ab initio* calculations of light nuclei in a harmonic oscillator basis. Phys. Rev. C **86**, 054002 (2012)
3. A.F. Lisetskiy, M.K.G. Kruse, B.R. Barrett, P. Navratil, I. Stetcu, J.P. Vary, Effective operators from exact many-body renormalization. Phys. Rev. C **80**, 024315 (2009)
4. M.K.G. Kruse, H.G. Miller, A.R. Plastino, A. Plastino, S. Fujita, Landau-ginzburg method applied to finite fermion systems: Pairing in nuclei. Eur. Phys. J. A Hadrons Nuclei **25**(3), 339–344 (2005)

Zeitfracht Medien GmbH
Ferdinand-Jühlke-Straße 7
99095 Erfurt, Deutschland
produktsicherheit@kolibri360.de